The Institute of Biology's
Studies in Biology no. 92

Nitrogen Fixation

John Postgate
F.R.S.
Professor of Microbiology and Assistant Director of the
A.R.C. Unit of Nitrogen Fixation, University of Sussex

EX LIBRIS

First published 1978
by Edward Arnold (Publishers) Limited
25 Hill Street, London W1X 8LL

Boards edition ISBN: 0 7131 2687 6
Paper edition ISBN: 0 7131 2688 4

Printed in Great Britain by
The Camelot Press Ltd, Southampton

General Preface to the Series

It is no longer possible for one textbook to cover the whole field of Biology and to remain sufficiently up-to-date. At the same time teachers and students at school, college or university need to keep abreast of recent trends and know where the most significant developments are taking place.

To meet the need for this progressive approach the Institute of Biology has for some years sponsored this series of booklets dealing with subjects specially selected by a panel of editors. The enthusiastic acceptance of the series by teachers and students at school, college and university shows the usefulness of the books in providing a clear and up-to-date coverage of topics, particularly in areas of research and changing views.

Among features of the series are the attention given to methods, the inclusion of a selected list of books for further reading and, wherever possible, suggestions for practical work.

Readers' comments will be welcomed by the author or the Education Officer of the Institute.

1978

The Institute of Biology,
41 Queens Gate,
London, SW7 5HU

Preface

The fixation of nitrogen—the conversion of atmospheric nitrogen to a form which plants can use—is a process which is fundamental to world agriculture. Though it has been recognized for over a century, scientific advances over the last two decades have radically altered understanding of the process. This booklet will survey the subject, with emphasis on recent developments, in an elementary manner. The background—the place of nitrogen fixation in the global nitrogen cycle—will be covered as well as recent developments in the biochemistry, physiology, genetics and ecology of nitrogen fixation.

1978   J. R. P.

Contents

1 The Nitrogen Cycle

1.1 Biological cycles

Living things, on this planet, continually recycle the chemical elements of which they are composed. The most familiar, day-to-day, example of such a cycling process is the carbon cycle. In this cycle, in its simplest form, green plants make use of photosynthesis to convert CO_2 from the atmosphere into plant material; animals, by consuming the plants, digesting them and using their carbon compounds for respiration, return the carbon atoms to the air as CO_2. Thus the carbon atoms of living things are continually recycled by way of the CO_2 in the air. The carbon cycle has many subtleties and sophistications when considered in detail but in crude outline the description just given is true: living things, on this planet, depend on the cycling of carbon between living matter and atmospheric CO_2, a process powered by the solar energy trapped by photosynthetic organisms. The major biological elements, carbon, nitrogen, oxygen and sulphur, are subject to comparable cyclic processes, and the most important, from both ecological and economic viewpoints, is the nitrogen cycle.

The nitrogen cycle can be drawn in several ways, of which that in Fig. 1–1 is a simple example. Essentially it symbolizes the transformations undergone by the element nitrogen (N) on this planet through the agency of living things. The left branch and lower sector show the synthesis of nitrogenous living matter (principally protein) from inorganic nitrogen compounds (nitrate, nitrite and ammonium ions) during growth of plants and their consumption by animals, followed by their return to the soil as a result of decay and putrefaction of plant and animal material. The upper sector shows the loss of nitrogen to the atmosphere from nitrates, and its return to the cycle by the process known as nitrogen fixation. The steps of the cycle will be discussed individually in a later section of this chapter; an important general point is that, in nearly all agricultural areas of this planet (areas where shortages of sunlight and/or water supply do no limit crop growth), biological productivity is determined by the availability of inorganic nitrogen in the soil. This means that in all but highly sophisticated agricultural communities, the rate at which the cycle turns determines biological productivity.

The element nitrogen is an essential constituent of all living things—the proteins and nucleic acids are their major nitrogenous constituents but most other biological materials contain some nitrogen atoms—and one can calculate that the plants and animals of the soils and waters of this planet together contain roughly 1.5×10^{10} tonnes of N. Each

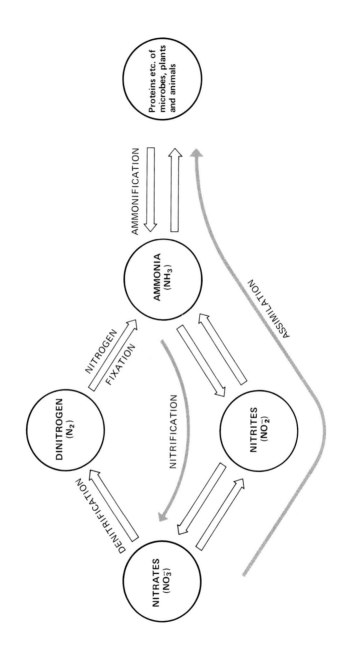

Fig. 1–1   The biological nitrogen cycle.

year the nitrogen cycle transforms roughly one-fifteenth of this (about 10^9 tonnes N; this number is difficult to calculate and may be out by a factor of 2 or more). Much of the cycling is by way of the lower part of the cycle, but roughly 2×10^8 tonnes N per year (a more reliable number by virtue of the acetylene test described in section 2.2) passes through the upper cycle. In practice, this means that input of inorganic N to the biosphere by the process called nitrogen fixation is rate-limiting for biological productivity on most areas of this planet's sea or land surface: only in undisturbed localities such as virgin savannah, or farmlands where N supplied by putrefaction processes or application of excessive nitrogenous fertilizer exceeds the N demand of the crops, do other nutrients (K, P, S or, rarely, a micronutrient such as Mg or Mo) become limiting.

1.2   Nomenclature

The term 'nitrogen' really refers to the N atom, and the gas which composes 80% of our atmosphere, N_2, is correctly termed dinitrogen. This term will one day become universally adopted but, since the more colloquial 'nitrogen' is still widely used for the gaseous element N_2, particularly in discussing the process of nitrogen fixation, I shall continue to use it in that context. 'Fixed nitrogen' will refer to any compound of N that is not N_2 but may have originated therefrom. Other technical terms will be explained as they arise; those which specially concern the nitrogen cycle are discussed in the next five sections.

1.3   Assimilation of nitrogen

This is a general term for the biological conversion of fixed nitrogen compounds, illustrated as nitrate, nitrite or ammonia in Fig. 1–1, to organic nitrogen. A growing plant or microbe in soil assimilates nitrogen as nitrate, converting it into protein, nucleic acids and minor nitrogenous components of the cell; an animal assimilates amino acids, building these into protein and other biological polymers. The pathways of nitrogen assimilation in living things together constitute a complex subject which could easily occupy a booklet of this size on its own. For present purposes, discussion will be restricted to the compounds illustrated in the cycle.

All plants and many bacteria reduce nitrates to ammonia by way of nitrites; the ammonia is then incorporated into nitrogenous biopolymers. This process is called *assimilatory* nitrate reduction, because the nitrogen of nitrate is assimilated into protein. It is the major process whereby nitrogen is incorporated into plant, and hence animal, material. The ammonia formed is assimilated primarily by incorporation into amino acids, and the commonest reaction of this kind is formation of glutamic acid from 2-oxoglutaric acid, a normal intermediate in carbon metabolism. The enzyme responsible for the reaction is called glutamate dehydrogenase and the reaction (1) is reversible:

$$NADH + HOOC.CH_2.CH_2.CO.COOH + NH_3 \rightleftharpoons$$
2-oxoglutarate   ammonia

$$HOOC.CH_2.CH_2.CHNH_2.COOH + H_2O + NAD \qquad (1)$$
glutamate   water

(NADH and NAD are the reduced and oxidized form of a biological reducing agent).

Glutamate is a key substance in the formation of other amino acids and of protein. An important point about this assimilation step is that the enzyme is not only reversible but has rather a low affinity for ammonia. Put another way, glutamate dehydrogenase is rather inefficient at picking up the ammonia molecule. Bacteria are frequently obliged to exist in conditions in which there is a shortage of ammonia, as well as of precursors such as nitrite or nitrate. In such circumstances, many can use a different series of reactions to assimilate the little ammonia available, one which makes use of two enzymes. Essentially, the first enzyme, glutamine synthetase, binds ammonia by reaction (2) to an existing molecule of glutamate to form glutamine:

$$HOOC.CH_2.CH_2.CHNH_2.COOH + NH_3 + ATP \longrightarrow$$
glutamate

$$NH_2.CO.CH_2.CH_2.CHNH_2.COOH + H_2O + ADP + P \qquad (2)$$
glutamine

(ATP and ADP are the charged and discharged forms of molecules responsible for the transfer of biological energy, inorganic phosphate (P) is released in the conversion of ATP to ADP, see section 2.2).

Glutamine synthetase is very efficient at picking up ammonia, but has the disadvantage as far as the economy of the microbe is concerned that one molecule of ATP, the biological 'quantum' of energy, is lost for each molecule of glutamine formed. Put in its simplest terms, the organism has to use up energy, as ATP, to make glutamine synthetase work. The glutamine so formed then reacts with 2-oxoglutarate, with the aid of a second enzyme, glutamate synthase, to form two glutamate molecules as in reaction (3).

$$NADH + NH_2.CO.CH_2.CH_2.CHNH_2.COOH$$
glutamine

$$+ HOOC.CH_2.CH_2.CO.COOH \longrightarrow$$
2-oxoglutarate

$$2HOOC.CH_2.CH_2.CHNH_2.COOH + NAD \quad (3) \qquad (3)$$
glutamate

By reactions (2) and (3) then the organism converts one glutamate molecule into two, assimilating ammonia efficiently but expending ATP.

Thus microbes are 'canny': they have two pathways of ammonia assimilation, one 'cheap' in terms of ATP but inefficient, for use when ammonia (or a precursor such as nitrate) is plentiful, and another, expensive but highly efficient, for use when supplies of ammonia are short. Once glutamate has been formed, the inorganic nitrogen of nitrate or ammonia has entered an organic compound on its way to being built into a living organism. Other routes exist whereby N can enter organic combination, but glutamate formation is the principal reaction. The enzymes of glutamate formation are also involved in regulatory processes (see section 6.3).

1.4   Ammonification

Inorganic nitrogen is returned to the cycle from organic matter as a result of autolysis, decay and putrefaction of biological material, and the principal form in which it appears is as ammonia. The general process is referred to as ammonification. The enzyme glutamate dehydrogenase, mentioned in the previous section, when working in reverse is typical of the enzymes responsible for this process; as in reaction (4) it forms a keto acid from an amino acid:

$$NAD + HOOC.CH_2.CH_2.CHNH_2COOH + H_2O \rightleftharpoons$$
$$\text{glutamate}$$
$$HOOC.CH_2.CH_2.CO.COOH + NH_3 + NADH \qquad (4)$$
$$\text{2-oxoglutarate}$$

A moment's inspection will show that reaction (4) is the reverse of reaction (1). Ammonia can also be released from amino acids by a hydrolytic reaction (the enzymes responsible are called deaminases) and from amides by deamidases; a familiar one is urease, which hydrolyses the urea of urine, causing the familiar smell of animal houses and imperfectly cared-for toilets. Details are inappropriate here, the essential point being that breakdown of proteins and other nitrogenous organic matter almost always leads to the formation of free ammonia.

1.5   Nitrification

The ammonification step of the cycle contributes free ammonia to the biosphere. Two classes of bacteria exist which live by oxidizing the ammonia to nitrate. Both are unusual forms of life: just as plants use sunlight to convert CO_2 to organic matter, these bacteria use the energy of ammonia oxidation to 'fix' CO_2. Like plants, they cannot use external sources of organic matter (such as glucose): for reasons which microbiologists still do not wholly understand, they are obliged to make their own organic matter from CO_2. Organisms of this kind are called *autotrophs*. Plants belong to the class of autotrophs called *phototrophs*, because they use light energy to fix CO_2; these bacteria are called *chemotrophs*, because they couple a chemical reaction to CO_2 fixation.

The biological oxidation of ammonia to nitrites and nitrates is called nitrification and the microbes responsible are called nitrifying bacteria. They can be divided into two main groups. *Nitrosomonas* is typical of six or so genera which oxidize ammonia to nitrite (see Fig. 1–1); *Nitrobacter* represents the rather fewer genera which oxidize nitrite to nitrate. Both classes of bacteria are very widespread but difficult to isolate and culture in the laboratory. From an ecological point of view they have two important functions, one beneficial and one deleterious. Ammonia, though an adequate plant nutrient, is less effective than nitrate and most plants prefer nitrate. The nitrifying bacteria perform a valuable function in rendering the nitrogen of ammonia more readily available to plants. On the debit side, ammonia is well retained by soil, whereas nitrates are readily washed away, so nitrification can result in loss of nitrogen from that zone of soil which is readily accessible to the roots of plants: the rain-washed surface layers. An example in which this process is economically important is the 'run off' of artificial fertilizer. Much of the artificial fertilizer used in agriculture is free ammonia or an ammonium salt; nitrification can lead to loss of nitrogen from the fertilized farmland to neighbouring soils and waters, occasionally causing contamination of drinking water. Ammonia from natural ammonification processes (e.g. a compost heap rich in nitrogen) can also become converted to nitrate and cause 'run off' problems; agricultural chemicals are available to restrict this process by inhibiting growth of nitrifying bacteria.

1.6   Denitrification

The process of assimilatory nitrate reduction was mentioned in section 1.3. Some microbes can use nitrates as a substitute for oxygen in respiration. The process is called *dissimilatory* nitrate reduction, because it is associated with the dissimilation (oxidation) of organic matter. Often the nitrate is converted to nitrite (advantage is taken of this reaction in the curing of bacon: the nitrite formed from nitrate reacts with meat protein, turning it red, altering its taste and protecting it from decay). Much of the nitrite formed then enters the assimilatory pathway of the nitrogen cycle (Fig. 1–1). There exist, however, several groups of bacteria which reduce nitrate to nitrogen gas (dinitrogen), and these bacteria are common in such environments as sewage treatment plants, compost heaps and rich soil. Examples are found in the genera *Pseudomonas*, *Micrococcus* and *Thiobacillus*. The dinitrogen is released and becomes part of the atmosphere. This process amounts to a net loss of biological nitrogen from the world's soils and waters; reliable data for the global scale of this process are not available, but a figure in the region of 2×10^8 tonnes N per year lost as N_2 is probably reasonable.

1.7   Nitrogen fixation

The loss of nitrogen to the atmosphere recorded above is substantial and, even if the number quoted is an order of magnitude out, it would soon bring life on this planet to a stop as a result of N-starvation. In fact, the fixation of nitrogen is the process which compensates for the net loss incurred by denitrification and, as I pointed out in section 1.1, the nitrogen economy of this planet happens to be such that, in most fertile areas, the rate of nitrogen fixation determines biological productivity.

The ability to fix nitrogen is restricted to the most primitive living things, the bacteria, and even within this group the property is by no means universal. The first product of fixation is ammonia, as the cycle in Fig. 1–1 illustrates, but it is important to emphasize that the ammonia is nearly always assimilated as rapidly as it is formed. From an ecological (or agricultural) viewpoint, the most important nitrogen fixers are those which fix in association with a plant (see Chapter 5), because fixed nitrogen is supplied just where it is needed: close to the plant's roots. As I stated in section 1.1, an annual input of some 2×10^8 tonnes of N takes place on this planet per year, and biological processes contribute the bulk of this.

Nitrogenous fertilizer is today mostly made from atmospheric dinitrogen by a reaction (5) which is known as the Haber–Bosch process, essentially a catalytic reduction of dinitrogen to ammonia:

$$N_2 + 3H_2 \;\rightarrow\; 2NH_3 \qquad\qquad (5)$$

The hydrogen is generated from natural gas and the reaction requires high pressures and moderately high temperature to be at all efficient. The world's industrial output of nitrogenous fertilizer increases annually and around 1975 was in the region of 3×10^7 tonnes N: about 15% of the total global N-input. Lightning and ultraviolet radiation cause the formation of nitrogen oxides, particularly in the upper atmosphere, and combustion (particularly the internal combustion engine) and electrical sparking generate oxides of nitrogen at ground level (in cities, the nitrogen oxides from car exhausts can be a serious atmospheric pollutant). These gases becomes washed into soil by rain and dew, contributing to the total N-input from atmospheric dinitrogen. Figures for this non-microbial input are not easily obtained, but authorities agree that such processes are unlikely to contribute more than 10%. It is reasonable, therefore, to accept that well over 70% of the input of the world's soil and water nitrogen is today supplied by nitrogen-fixing bacteria in one form or another.

It is important to put this matter into perspective. The amount of N present on this planet as dinitrogen is vast; something like 4×10^9 tonnes in the atmosphere and a surprising 2×10^{11} tonnes bound in sedimentary and primary rocks beneath the surface. None is accessible to plants until it is fixed, principally by microbes; in many circumstances the microbes have to die and decompose—ammonification has to take place—before the plants can gain access to the nitrogen fixed. In symbiotic systems, in which nitrogen-fixing bacteria are intimately associated with plants, the transfer of fixed nitrogen from microbe to host is much more rapid and efficient, so symbioses of the kind discussed in Chapter 5 are by far the most important agents of biological fixation as far as global biological (and agricultural) productivity are concerned. None the less, free-living microbes make a serious contribution, particularly in poor or unfertilized soils. Many blue-green algae fix nitrogen and can be the primary source of N-input in the tundra, in the sea, on stony or devastated land. This matter is discussed further in Chapter 4. In highly developed agricultural countries, where ammonia fertilizer is available and cheap, one might expect the biological processes to be unimportant. Yet agronomists have evidence, indicated in Table 1, that only about one-third of the nitrogen

Table 1   Estimated N-input into plant crops in 1971–2

| Source of N | millions of tonnes of N | | | |
	UK	USA	Australia	India
Fixation in legumes	0.4	8.6	12.8	0.9
Fixation by free-living microbes	<0.05	1.4	1.0	0.7
Taken up from N fertilizer	0.6	4.9	0.1	1.2

produced, in plants, by U.S. agriculture in 1971–2 entered the soil as artificial fertilizer; the rest appeared by biological fixation. In Australia, where the exploitation of nitrogen-fixing plants in agriculture is highly developed, chemical fertilizer accounted for less than 1% of crop nitrogen. It is curious that, in India, where soil and crop yields are poor yet demand is intense, about 40% of the crop nitrogen nevertheless originated from industrial fertilizer. The possibilities of exploiting the biological process to alleviate the near-starvation condition in much of India are obvious.

The principal nitrogen-fixing systems useful in world agriculture are the legumes: peas, beans, soybeans, chickpeas, lupins, lucerne and so on. They all involve association of a species of the bacterial genus *Rhizobium* with a plant, and characteristically the bacteria colonize zones of the roots in little excrescences called nodules (see Chapter 5 and Fig. 5–1). Such plants can be made use of in three principal ways. Direct use as food is most familiar: peas and beans are a well known and regular part of most people's diet; rather less well known is the now widespread use of protein from soybeans in the food industry. Conversion to meat products via cattle fodder is the second use: pulses and lucerne (alfalfa) are increasingly used as cattle food supplements, both as the plant and after conversion to silage. The third major use is as green manure: clovers, melilot and lucerne are grown and ploughed in to up-grade poor or spent agricultural land, particularly in communities which practise intelligent crop rotation. Use of clover to up-grade the fertility of Australian soil led to dramatic improvements in the 1940s. Sometimes a given plant system can serve two purposes: lupins have been recommended for agricultural use in Western Australia because their seeds prove to be a useful high protein grain for animal feed and the plants can be ploughed in as a green manure after harvest.

1.8   Nitrogen fixation and energy resources

An important feature, particularly to poor countries, of biological nitrogen fixation is that the energy for the conversion of dinitrogen to ammonia generally comes from sunlight. Legumes, for example, supply the products of photosynthesis to the bacteria in their nodules, who use them for fixation. Blue-green algae use sunlight both to grow and fix nitrogen. In sharp contrast, the Haber process requires considerable energy to prepare hydrogen from natural gas and rather less energy to compress the reagents and heat them. In addition, quite a complex industrial installation is required, so production must be localized, and a further consumption of energy (as fuel for lorries, trains, tractors and so on) is required to transport the ammonia from the factory to where it is needed. Transport is far from a trivial question: in 1964, well before the 'energy crisis', scientists calculated that transport facilities equivalent to the world's shipping tonnage at that time would be needed by the year 2000, solely to transport fixed nitrogen from factories to farms, if the rate of population growth at that time was sustained. The rate of population growth has not changed significantly, in global terms, but it is now clear that the world's energy supplies must be conserved. Petroleum products, particularly, are escalating in cost. Petroleum costs do not seriously affect production costs of ammonia fertilizer, because the primary energy source is methane, but they affect secondary costs such as packaging and transport, and costs of natural gas tend to rise with other energy costs. A dramatic rise in the costs of ammonia fertilizer took place in the early 1970s (4.5-fold between 1972 and 1975) but, though energy costs played a small part, the major cause was a steadily increasing world demand for fertilizer which outstripped supply. More factories can be built, and more ammonia fertilizer can be made, but in the long run the energy costs for methane, transport and so on are likely to become prohibitive. The global need to conserve the fossil fuels, coal, methane and petroleum, provides as important a reason for the exploitation of biological nitrogen fixation as does the escalating demand for food.

2 The Enzyme

2.1 Nitrogenase

Nitrogen-fixing bacteria conduct a reaction which, in chemical terms, is most unusual. The dinitrogen (N_2) molecule is normally very unreactive: red hot magnesium, or a catalyst at elevated pressures and temperatures as in the Haber–Bosch process, are normally necessary to make N_2 reactive. In part this chemical inertia is due to the very stable triple bond which links the two atoms in the structure $N \equiv N$. Nitrogen-fixing organisms make the N_2 molecule reactive at ordinary temperatures and pressures and, in addition, they do so in water under an atmosphere of oxygen—both substances which would interfere drastically with such reagents as red hot magnesium or the Haber system! The biological catalyst, or enzyme, responsible for this process must have some exceptional chemical properties, and the way in which it functions is a fascinating problem which has only partly been resolved.

The enzyme was given the name 'nitrogenase' long before any such substance could be isolated. Though the name does not conform to systematic enzyme nomenclature, it is easily understood and useful. Many attempts were made, in the 1940s and 1950s, to extract nitrogenase from nitrogen-fixing bacteria but consistent success was only achieved when, in 1960, a group of workers in the Central Research laboratories of Dupont de Nemours (a large U.S.A. chemical firm) obtained a solution of nitrogenase from *Clostridium pasteurianum*. *C. pasteurianum* is an anaerobic bacterium common in soil and compost heaps; it is unable to grow in air yet was one of the first nitrogen fixers to be discovered. The secret of the Dupont group's success was the fact that, because the organism was anaerobic, they excluded air carefully from the extract at all times. The extract was a sort of juice prepared by resuspending dried bacteria in aqueous salt solutions and centrifuging, so it was not too difficult to keep air away in the laboratory. The extract converted N_2 to ammonia (NH_3), a reaction proved by using the isotope of nitrogen $^{15}N_2$ and isolating $^{15}NH_3$. The critical point, which later proved to have an importance far beyond the isolation of the enzyme, is that the activity was completely and rapidly destroyed if the extract was exposed to air.

Once this secret was uncovered, progress was rapid and, by 1974, nitrogenases had been extracted from some twenty-five nitrogen-fixing microbes, including *Rhizobium* after it had colonized the roots of leguminous plants. More impressive, perhaps, is the fact that biochemists had developed ways of handling enzymes in the absence of air, excluding air from electrophoresis systems, chromatography columns, centrifuges and so on, and had actually purified nitrogenase from about five microbes. The enzymes proved to have some remarkable properties, not least that nitrogenases proved to be extremely similar from very different sources.

Table 2 gives some properties of the enzyme purified from three different species of nitrogen-fixing bacteria. *C. pasteurianum* is the anaerobe from which the first extract was made; *Azotobacter chroococcum* is a very different microbe, which can only grow in air and which is widespread in soils, particularly chalky or sandy ones; *Klebsiella pneumoniae* is physiologically intermediate but not related to either: it can grow either with or without air but it can only fix nitrogen in the absence of air. *K. pneumoniae*, too, is a common soil bacterium but it is often present in mammalian intestines (and, occasionally, in a rare lung disease from which its name derives).

The first thing that Table 2 shows is that the enzyme is not one protein. On purification it separates into two absolutely distinct proteins, both dark brown in colour, neither active without the other. Both are complex proteins: the big one consists of four sub-units (polypeptide chains) stuck together in some way, two of the sub-units being different from the other two. The smaller protein consists of only two sub-units, apparently identical. Both contain iron atoms, the amount per molecule of the large protein being substantial but rather variable; the smaller one having four per molecule. In both cases the iron is accompanied by essentially the same number of sulphur atoms. (These sulphur atoms are termed labile sulphur in the table; this means sulphur which can be displaced as H_2S by weak mineral acid. It is known to be sulphur actually attached to the iron atoms.) The big protein also has two atoms of the metal molybdenum per molecule.

(Iron and molybdenum belong to the class of elements which chemists call the 'transition metals', a rather ill-defined class including group VIII of the periodic table and some in groups VII and VI.)

A second point shown by Table 2 is that the two proteins are very much alike no matter which microbe they come from. Their similarity is so great that, in the laboratory, one can take the big protein from *Klebsiella* (Kp1) and mix it with the small protein from *Azotobacter* (Ac2) and get a fully active 'hybrid' nitrogenase enzyme. The converse mixture (Kp2 + Ac1) is also fully active, but crosses with the clostridial proteins give only weakly, if at all, active enzymes.

Most impressive is the readiness with which the proteins are destroyed by oxygen, particularly the smaller protein. This is true even when the proteins are purified from *A. chroococcum* which, because of its aerobic habit, might have been expected to have more oxygen-tolerant nitrogenase. In fact, there is a sense in which it has, because crude extracts of *A. chroococcum* (made by squashing the bacteria in an appropriate apparatus and removing the cell envelopes) contain the nitrogenase in

Table 2 Some properties of the two component proteins of nitrogenase purified from three species of bacteria

Protein	Cp1	Kp1	Ac1	Cp2	Kp2	Ac2
Molecular weight (mw)	220 000	218 000	227 000	55 000	66 700	64 000
Number and mw of sub-units	$2 \times 50\,700$ + $2 \times 59\,500$	$2 \times 51\,300$ + $2 \times 59\,600$	probably 2×2 types	$2 \times 27\,500$	$2 \times 34\,000$	$2 \times 31\,000$
Mo atoms/molecule*	2	2	2	0	0	0
Fe atoms/molecule*	22–24	30–35	21–25	4	4	4
Sulphide groups/molecule*	22–24	>18	18–22	4	4	4
Half life in air (minutes)	short	10	10	very short	0.75	0.5

Cp means *Clostridium pasteurianum*, Kp means *Klebsiella pneumoniae*, Ac means *Azotobacter chroococcum*. 1 or 2 refer to the larger and smaller proteins irrespective of origin.

* Data rounded to nearest whole numbers.

small, sub-cellular particles in which the proteins are quite tolerant of oxygen. Only when the particle is broken up and the proteins purified from it do they show the high oxygen sensitivity illustrated in Table 2.

2.2   Substrates for nitrogenase

Nitrogenase, then, is a binary enzyme, consisting of two proteins. It reduces nitrogen to ammonia and this, with the purified enzyme, is the only detectable product. One molecule of N_2 is converted quantitatively to two molecules of NH_3. One might expect intermediates such as the rather unstable compound di-imide $(HN=NH)$ or the more stable hydrazine $(H_2N—NH_2)$ to be formed as intermediates, but careful tests for these compounds using the isotope $^{15}N_2$ have given no sign of them. The conversion of N_2 to NH_3 is a reduction (like the Haber–Bosch process) and a reducing agent corresponding to the hydrogen of the Haber–Bosch process is needed. The nature of this reducing agent is discussed further in the next chapter; in the earliest experiments it was present in the cell extracts. Research on the enzyme was greatly facilitated by the discovery, in 1964, that a simple laboratory chemical would substitute for the reducing agents normally present in bacteria. This substance is sodium dithionite (earlier known as sodium hydrosulfite), $Na_2S_2O_4$, and it is now very widely used as a reductant in enzyme studies.

A feature of nitrogenase is that, in order to function at all, it needs adenosine triphosphate (ATP). This substance is normally present in all living tissue, and is the principal molecule whereby chemical energy is mobilized for biological processes. ATP is hydrolysed to adenosine diphosphate (ADP) with a net loss of chemical energy which can be used by living cells for all sorts of biological processes; when nitrogenase functions it converts a substantial amount of ATP to ADP, the energy released being somehow used in the fixation process. In the test tube, some 12 to 15 ATP molecules are consumed to convert one N_2 to two NH_3 molecules. In physiological terms, nitrogen fixation is an expensive process, because the organism must use food (glucose for example) to make that ATP; it usually regenerates ATP from ADP by processes which are well understood but which need not concern us here. Historically, the involvement of ATP took rather a long time to discover, because of another unexpected property of the enzyme. If you just add ATP, the reaction starts quite well but soon stops, and the reason is that ADP is an inhibitor: if ADP is allowed to accumulate, it blocks some step in the reaction and everything comes to a standstill. Research workers today usually use an artificial ATP-regenerating system to overcome this problem.

Another essential ingredient for nitrogenase is the metal magnesium, as its ion, Mg^{2+}. If this is completely absent, the enzyme will not function. Quite what the Mg^{2+} is doing is still not understood, but one fact is already clear: ATP does not react with nitrogenase alone; it has to become the mono-magnesium salt (often written MgATP, though this is in no sense a correct chemical formula) to perform its function with nitrogenase.

Nitrogen fixation therefore requires rather a complicated system: two proteins, a reducing agent, ATP and Mg^{2+}. And the complications are not over yet. If dinitrogen is present, it is reduced to ammonia. But if it is not present (with an active preparation under argon, for example), the enzyme then reacts with water, forming gaseous hydrogen. What happens is that hydrogen ions (H^+) of water become reduced and join together to form H_2 molecules: the enzyme is a powerful enough reducing agent to decompose water. In fact, this reaction actually accompanies nitrogen fixation when the enzyme is used in the test tube and there is evidence that it takes place in living bacteria, but that they have ways of recycling the hydrogen formed. The evolution of hydrogen requires a reducing agent, ATP and Mg^{2+}, of course, just like the nitrogen fixation reaction proper.

Nitrogenase reduces the triply bonded $N \equiv N$ molecule to ammonia. A number of molecules exist which are a little like dinitrogen in that they are small and have a triple bond. Examples are:

H	H	H	N	C$^-$
|	|	|	|||	|||
C	C	N$^-$	N$^+$	O$^+$
|||	|||	|	|	
C	N	N$^+$	O$^-$	
|		|||		
H		N		
acetylene	hydrogen cyanide	hydrogen azide	nitrous oxide	carbon monoxide

The first four substances are indeed reduced by nitrogenase, provided ATP and Mg^{2+} are available, but the fifth, carbon monoxide, is not reduced. However, carbon monoxide certainly interacts with the enzyme since it blocks the reduction of all the others, and of nitrogen, if it is present. Therefore it is thought to bind to whatever the site on the enzyme is which takes up N_2, but to be incapable of being reduced. It is interesting, however, that carbon monoxide does not interfere with the hydrogen evolution reaction mentioned in the previous paragraph.

Acetylene is a particularly important substrate because the product, ethylene, can be detected rapidly and with great sensitivity by gas chromatography. This reaction is the basis of the now famous acetylene test for nitrogen fixation, because it takes place not only with enzyme preparations but also with living bacteria, plants or excised nodules, soil or water samples. One has only to expose a nitrogen-fixing system to acetylene, in appropriate conditions of temperature, oxygen content and so on, and, sometimes within minutes, the rate of ethylene formation will provide a measure, usually quantitative, of the nitrogen-fixing capacity of the system under test. This test works with almost all biological nitrogen-fixing systems, and its application has not only revolutionized the study of nitrogen fixation in nature, but permitted the rapid progress in understanding the biochemistry and genetics of nitrogen fixation which took place in the 1970s.

A substrate whose reduction was of some scientific importance when it was discovered is methyl isocyanide. This compound can be reduced by ordinary chemical reducing agents such as sodium borohydride, and the product is dimethylamine:

$$CH_3NC \xrightarrow{NaBH_4} CH_3NHCH_3 \qquad (6)$$

methyl isocyanide   dimethylamine

Nitrogenase gives different products, apparently splitting the NC bond so that methylamine and methane are formed:

$$CH_3NC \xrightarrow{nitrogenase} CH_3NH_2 \ + \ CH_4 \qquad (7)$$

methyl isocyanide   methylamine   methane

The enzyme also gave some unexpected by-products (ethane and acetylene) along with methane. The enzymic kind of reaction was then found to occur with sodium borohydride, but only if the methyl isocyanide had first reacted with a transition metal (such as platinum) to form a complex. These observations suggested that, in the enzyme, isocyanide and, by implication, comparable substrates such as nitrogen and acetylene, became bound to one of the transition metal atoms in the protein (iron or molybdenum) before becoming reduced. The reaction of nitrogenase with isocyanide is an example of the sort of experiment, making use of 'unnatural' substrates for the enzyme, which has given information about the details of the way in which the enzyme attacks its substrate.

Nitrogenase is thus a very complicated enzyme with several unusual properties which are summarized briefly below:

1. consists of two proteins;
2. is destroyed by oxygen;
3. contains the transition metal atoms iron and molybdenum;
4. needs Mg ions to be active;
5. converts ATP to ADP when functioning;
6. is inhibited by ADP;
7. reduces N_2 and several other small, triply bonded molecules;
8. reduces hydrogen ions to gaseous hydrogen.

Even this list of properties shows that the major question of how nitrogenase works needs answers to a number of subsidiary questions. For example, how and in what proportion do the two proteins interact? Why and where does ATP react? Why are the metal atoms there? Where do dinitrogen or other substrates bind? Is hydrogen evolved from the same or a different site? Why are no partially reduced intermediates formed from N_2? Present knowledge cannot completely answer any of these questions, but the study of nitrogenase has advanced sufficiently to allow at least a provisional idea of the way it functions.

2.3 The two proteins and the nitrogenase complex

Nitrogenase is rich in iron, and it is probably this element which is responsible for the brown colour of the two proteins. In other proteins which contain iron, such as the haemoglobin of blood, conventional visible and ultraviolet spectroscopy gave invaluable information about its behaviour. The spectra of nitrogenase proteins are rather featureless, however, and scope in this direction has been rather limited. There exists, however, a form of spectroscopy (Mössbauer spectroscopy) which uses γ-radiation instead of light and which works very well with chemicals containing iron. Such spectra of nitrogenase proteins were particularly revealing as far as the molybdoprotein is concerned. They showed that this protein could exist in three different states, distinguished by the proportion of the iron atoms which were in the reduced (ferrous) condition in the molecule. The most oxidized form is probably physiologically unimportant; the intermediate form is that in which the protein is usually isolated; the most reduced form only appears when the other protein is present together with ATP, Mg^{2+} and sodium dithionite. In other words, the most reduced form appears only when all the

Fig. 2–1 A scheme for the action of nitrogenase. **P1** is the larger protein containing iron and molybdenum atoms, **P2** is the smaller protein with iron atoms only. **F** is an electron-donating substance (ferredoxin or flavodoxin) which donates an electron (a black dot) to iron atoms in **P2**. ATP reacts with magnesium ions to produce a compound which activates the reduced form of **P2**. Meanwhile **P1**, with a 'spare' electron among its iron atoms, has bound a reducible substrate such as N_2 at transition metal atoms depicted as molybdenum (see the text). Activated **P2** joins **P1** carrying substrate to form the 'nitrogenase complex', assisted by ATP-Mg, within which an electron from the iron atoms in **P2** is transferred to the iron atoms in **P1**, prior to reaching the bound substrate. After several such electron transfer events, the product, such as NH_3, of enzyme action is released. Each electron transfer requires a new electron from **P2** and each time transfer takes place a molecule of ADP is formed from the ATP.

The scheme is one of several which could be proposed based on the experimental findings that **P2** is concerned with ATP consumption and electron transfer, **P1** with substrate binding and reduction.

components necessary for functioning nitrogenase are supplied, and is, in fact, the predominant form in the functioning enzyme.

Proteins with iron in them have unusual magnetic properties. A very sensitive kind of spectroscopy which exploits these properties is called electron paramagnetic resonance, usually abbreviated to epr. The intermediate form of the molybdoprotein shows a very clear and distinct epr absorption (the other two forms do not), and the character of that absorption is altered in a subtle way by a substrate such as acetylene. This is a fragment of evidence suggesting that this particular protein reacts with substrates such as acetylene. Scientists, suspecting that a metal atom binds such substrates, are tempted to the view that the molybdenum atoms, rather than the iron atoms, bind the reducible substrates, but there is no firm evidence for this view.

epr spectra have been very useful in assigning a function to the smaller protein of nitrogenase. This can exist in oxidized or reduced forms; only the latter has an epr spectrum. An important observation is that, if Mg^{2+} and ATP are added, the epr spectrum of the reduced form undergoes a substantial change. This change is accompanied by an increase in chemical reactivity: the protein is even more easily destroyed by oxygen, loses its iron more easily, reacts with other reagents more rapidly. Effectively it becomes a more powerful reducing agent, one capable of generating the most reduced form of the molybdoprotein.

Evidence of this kind suggests that the molybdoprotein is involved in picking up and reducing substrates such as dinitrogen and acetylene and that the smaller protein reacts with ATP and generates the reducing form of the other protein. But do the proteins act in sequence or do they join together to form a complex which is nitrogenase? Studies on rates of reaction suggest that they join up to form a complex, but perhaps the most convincing evidence is that, if one mixes the oxidized form of the small protein with the intermediate form of the molybdoprotein, and then sediments them in an ultracentrifuge, they sediment together as a complex containing molecules of each protein.

In summary, then, present evidence is that, for nitrogenase to work, a complex of two proteins must be formed in which the molybdoprotein has the dinitrogen molecule bound, at some metal atom which may well be molybdenum, and the smaller protein has ATP bound (as its monomagnesium salt). The fixation process involves reduction of iron in the molybdoprotein at the expense of iron in the smaller protein; as a consequence ATP becomes converted to ADP. These reactions are brought together in a scheme in Fig. 2–1; it must be taken as suggestive rather than as representing established fact but it represents a substantial advance in understanding nitrogenase since the first extracts were obtained in 1960.

3 Physiology

3.1 The need to exclude oxygen

The biochemical properties of nitrogenase present the organism with a number of problems of a physiological character. Perhaps the overriding one arises from the sensitivity of the proteins themselves to oxygen: how, one can ask, do microbes fix nitrogen on a planet covered by 20% oxygen, when their enzyme is irreversibly destroyed on contact with that gas? The answer has proved very revealing. Nitrogen-fixing systems do not in fact function in the presence of oxygen, and the microbes which can fix nitrogen in air use various stratagems to exclude oxygen from the nitrogen-fixing site. When the organism is capable of anaerobic metabolism the stratagem is rarely complex, but obligate aerobic nitrogen-fixing bacteria have a more subtle problem because oxygen is essential for their general cell metabolism, without which no nitrogen could be fixed. So they are obliged to admit sufficient oxygen to sustain multiplication and ATP generation while not damaging nitrogenase. An even more awkward problem is faced by the blue-green algae (correctly known as blue-green bacteria) which are obliged to reconcile nitrogen fixation with the oxygen evolution step of photosynthesis. In this section the physiological processes whereby microbes protect nitrogenase from oxygen will be discussed in approximate order of increasing sophistication.

3.1.1 Avoidance

Nitrogen-fixing bacteria such as *Clostridium pasteurianum* or *Desulfovibrio desulfuricans* (the sulphate-reducing bacteria) are obligate anaerobes even when they are provided with fixed nitrogen, and they can be said to avoid the problem as far as possible. *D. desulfuricans*, in fact, shows no nitrogenase activity if a sample from a nitrogen-fixing culture is exposed to air before testing. This is not true, however, of *C. pasteurianum*. Though the cultures will only grow in the absence of air, the live cells can be handled in air, even centrifuged and transferred in air as a live paste; only when the cells are disrupted by some treatment is the nitrogenase destroyed by air.

3.1.2 Facultative nitrogen-fixing bacteria

Many bacteria are facultative anaerobes: capable of either aerobic or anaerobic growth. If such bacteria are also capable of fixing nitrogen, they are usually able to do this only in anaerobic conditions. Typical of this physiological type are nitrogen-fixing members of the genera

Klebsiella, *Enterobacter* and *Bacillus*. The photosynthetic bacterium *Rhodospirillum rubrum*, whose photosynthesis does *not* lead to oxygen evolution, is capable of growing aerobically but only non-photosynthetically in the dark. It is able to fix nitrogen but, logically enough, only when growing anaerobically in the light.

The oxygen relations of *Klebsiella pneumoniae* have been studied in some detail and the statement that it cannot fix nitrogen in the presence of air requires qualification. Provided the population of *Klebsiella* is sufficiently dense, as it is, for example, inside a colony on nutrient agar, some fixation of nitrogen can occur. The determining factor is whether the cells, by their collective respiration, can lower the oxygen tension in their own vicinity to zero. Healthy colonies which fix nitrogen can be grown, for example, on agar slopes if the medium is free of inorganic nitrogen but contains a limiting amount of organic nitrogen (such as $200 \mu g$ casamino acids/ml). Continuous cultures of *Klebsiella pneumoniae* can be established in a liquid variant of such a medium and will grow and fix nitrogen indefinitely under an atmosphere containing oxygen provided the culture is not stirred too vigorously (fast stirring increases the rate at which oxygen dissolves in water). An interesting observation is that, in such conditions, the organisms are actually growing aerobically: they are 'oxygen-limited' in the sense that they are consuming oxygen as fast as it is supplied. As will be explained in section 3.3, they are able to use products of their aerobic metabolism for nitrogen fixation. Physiologically, they are right on the threshold of being nitrogen-fixing aerobes.

If detectable dissolved oxygen *is* present in the culture fluid, *K. pneumoniae* does not fix nitrogen. A simple question thus arises: does it make nitrogenase only to have it destroyed by oxygen? Or does it cease to make the enzyme when oxygen is present? In fact it does the second. Oxygen-damaged nitrogenase in cells can be detected quite easily by immunology or by electrophoresis of extracts of the organisms, and experiments using both techniques indicate that no nitrogenase is formed in the presence of oxygen. Oxygen thus regulates biosynthesis of nitrogenase in *K. pneumoniae*: the organism has some sensing equipment which recognizes the presence of free oxygen and 'switches off' nitrogenase synthesis. The nature of this sensor is at present unknown.

3.1.3   Microaerophily

There exists a class of bacteria which is naturally sensitive to oxygen yet which can grow aerobically. An example is *Lactobacillus casei*, one of the normal inhabitants of milk, which grows in air but only if the partial pressure of oxygen is low. Such bacteria are common in soil and specialized habitats to which air has limited access. Many of the aerobic nitrogen-fixing bacteria behave like microaerophiles when fixing nitrogen in air, though in the absence of air they behave as perfectly normal aerobes. A well established example occurs with the tropical nitrogen-fixing aerobe *Derxia gummosa*. When a population of this organism is spread on nitrogen-deficient nutrient agar, a few of the organisms form massive, glutinous colonies which, the acetylene test shows, fix nitrogen vigorously. Those colonies have a yellowish tint. The remainder are small, white colonies which fix no nitrogen and which do not develop further (Fig. 3–1). This colonial dimorphism, as it is called, is

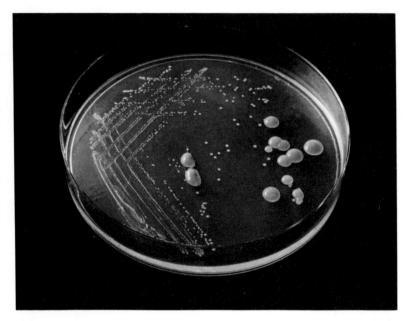

Fig. 3–1   *Derxia gummosa* colonies grown in air on a jellified medium free of fixed nitrogen. (Courtesy of Dr Susan Hill.)

very consistent and subcultures from either small or large colonies show a similar pattern. It can be altered either by incubating the plates at a low oxygen tension (10% O_2 instead of the 20% O_2 of ordinary air) or by adding a small amount of an ammonium salt to the agar. The explanation of the dimorphic growth in air seems to be that media set with agar contain small amounts of fixed nitrogen (probably as an impurity in the agar itself, which is a polysaccharide obtained from seaweed) and a few colonies 'scavenge' this and reach a critical size such that, at their centres, the oxygen tension is low enough for nitrogen fixation to be initiated. These become the large, glutinous, yellowish ones; the other colonies obtain insufficient fixed nitrogen to reach a critical mass and do not develop. A lowered oxygen tension decreases the 'critical mass', so that all colonies get started; a trace of ammonium salt provides sufficient scavengeable fixed nitrogen for them all to reach the critical mass.

The most recent (and, in its time, revolutionary) example of oxygen sensitivity is the demonstration that some strains of *Rhizobium* show intense microaerophily. Rhizobia grow readily in air in media containing fixed nitrogen such as glutamate or nitrate but, for many years, scientists believed that rhizobia could only grow and fix nitrogen symbiotically; that is to say, in intimate association with a plant. However, in 1975 scientists in Australia and Canada discovered that, at very low oxygen tensions, rhizobia belonging to the slow-growing 'cowpea' group would grow and fix nitrogen in the complete absence of plant material; this subject is discussed further in section 4.3.

Oxygen sensitivity amounting to microaerophily is shown by several other types of nitrogen-fixing aerobe, including *Mycobacterium flavum*, *Spirillum lipoferum*, *Corynebacterium autotrophicum* and the methane-oxidizing bacteria. They are discussed further in section 4.3.

3.1.4   Respiration

The facultative and microaerophilic nitrogen-fixing bacteria are able to fix nitrogen in air only if their respiration enables them to decrease the oxygen level of their environment to innocuous levels. In all aerobic living things, respiration is a means of generating biological energy (as ATP) from the energy of combustion of foodstuffs; a highly efficient means of energy generation compared with the fermentative processes which are characteristic of most anaerobes. In these nitrogen-fixing bacteria, respiration is performing a protective function, that of protecting the nitrogenase from oxygen damage. Certain of the azotobacters (such as *A. chroococcum*, see Fig. 3–2, or *A. vinelandii*) are very well adapted to aerobic nitrogen fixation and, though they do show oxygen sensitivity at high aeration levels or with hyperbaric oxygen, they have no difficulty in fixing nitrogen in air. A characteristic feature of such organisms is that they have exceptionally high respiration rates; 10 to 50 times the rate of *Mycobacterium flavum*, for example. In these organisms the protection of nitrogenase has become the dominant function of respiration: the organisms' respiration scavenges oxygen from the nitrogen-fixing site. This process is known as 'respiratory protection' and it is now well established in the context of aerobic nitrogen fixation. (Similar processes may well operate in the protection of other oxygen-sensitive enzymes in microbes, but they are less clearly established.) *A. chroococcum* consumes about 1 g of sugar (for example) to fix 10 to 15 mg of oxygen supply and a growing culture can double its respiration rate over about 20 minutes if the oxygen content of the atmosphere is doubled.

An obvious consequence of respiratory protection is that enormous amounts of carbon source must be consumed to scavenge the oxygen, so growth must be very inefficient in terms of the food consumed by the

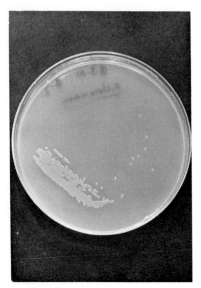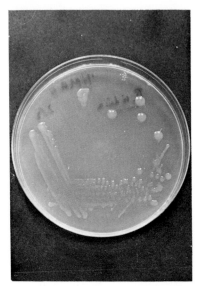

Fig. 3–2 Colonies of *Azotobacter chroococcum* (left) and *Beijerinckia indica* (right) growing on a jellified medium free of fixed nitrogen. Notice the glutinous quality of *B. indica* colonies.

bacteria. This in fact proves to be true. An average laboratory culture of *A. chroococcum* consumes about 1 g of sugar (for example) to fix 10 to 15 mg of N. If the population is starved of oxygen, so that little respiratory protection is necessary, up to 45 mg N can be fixed for each gram of sugar consumed. Conversely, if aeration rates are high, fixation becomes very inefficient and amounts lower than 1 mg N can be fixed per gram of sugar. Respiratory protection consumes a lot of substrate and, in nature, substrates are rarely available in great quantities.

A second consequence of respiratory protection concerns the generation of ATP. Although protection of nitrogenase is the dominant physiological function of *Azotobacter*'s normal aerobic respiration, the organism still has need of biological energy and ATP must be generated. But respiratory protection might thus lead to generation of a vast excess of ATP. In practice, the organism proves to be able to regulate its ATP production at least to some extent. Though it is unable to uncouple respiration from ATP generation completely, it possesses two distinct biochemical respiration pathways. One is of high efficiency, and is dominant when the organism is growing with a relatively restricted supply of oxygen; the other is of lower efficiency, generating about a third as much ATP for a given amount of substrate, and is dominant in conditions of high oxygenation. Logically enough, the low efficiency pathway is used when respiratory protection is of primary importance.

Respiratory protection has evolved to a relatively elaborate level in *Azotobacter*, and one can see a clear sequence of increasing physiological sophistication when one considers oxygen-scavenging in the facultative bacteria, the microaerophiles and *Azotobacter*. But respiratory protection is not the only oxygen-restricting process in *Azotobacter*; a back-up protection apparatus will be discussed in section 3.1.6.

3.1.5   Slime

Aerobic nitrogen-fixing bacteria very frequently form large quantities of extracellular polysaccharides: the glutinous quality of colonies of *Beijerinckia*, for example, can be quite spectacular (Fig. 3–2). Though not firmly established, it seems likely that this slime layer impedes diffusion of oxygen to the bacteria within and represents a crude physical barrier against excessive oxygenation.

3.1.6   Conformational protection

Crude cell-free extracts of *Azotobacter chroococcum* or *A. vinelandii* do not contain the nitrogenase protein as an oxygen-sensitive solution. The proteins are in a particle-like body which can be sedimented in an ultracentrifuge and within which the nitrogenase is relatively stable in air (Chapter 2). This particle includes proteins other than nitrogenase. Association with certain other proteins thus seems to decrease the oxygen sensitivity of nitrogenase and this observation has led to the idea of conformational protection. It seems that, in the oxygen-tolerant particle, the oxygen-sensitive sites of the nitrogenase proteins can assume some conformation (relative to each other or to other molecules) in which they are physically protected from oxygen damage. It is assumed that the enzyme cannot function in the conformationally protected state. The importance of this idea is that it can be extended to account for the response of *Azotobacter* to an oxygen stress too great for respiratory protection to cope with, for in these circumstances the organism ceases fixing nitrogen abruptly but the nitrogenase proteins remain undamaged and can, if exposure to oxygen stress is brief, resume fixation as soon as appropriate conditions of oxygenation are restored. It has been suggested that the nitrogenase in the living cells can reversibly assume the conformationally protected (but inactive) state represented by the sub-cellular particle, and that this is a protective stratagem which the organism can bring into play when its respiration is unable to cope with an oxygen stress. An implication of this view is that the enzyme is normally in a sub-cellular compartment in the living cell, close to protective proteins. While it is plausible, there is no substantial evidence for this view at present.

3.1.7   Heterocysts

Sub-cellular compartmentation in *Azotobacter* is still a matter of hypothesis. Compartmentation at a grosser level in blue-green algae is well established. Certain of the filamentous blue-green algae such as *Anabena* have, when they are fixing nitrogen, specialized cells called heterocysts at regular intervals along their filaments (Fig. 3–3). These do

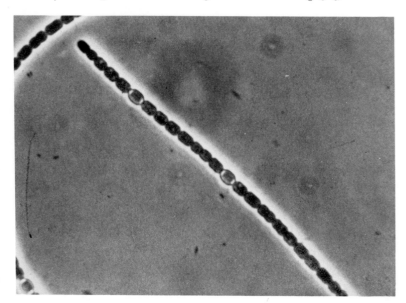

Fig. 3–3   A photomicrograph of *Anabena cylindrica* showing two enlarged cells (heterocysts) on the principal filaments. (Courtesy of Dr P. Fay.)

not appear when the algae are provided with adequate fixed nitrogen as ammonia (nitrate is less efficient in suppressing heterocyst formation). Substantial evidence has now accumulated to the effect that nitrogenase is normally restricted to the heterocysts and is absent from the vegetative filament. Correspondingly, the ordinary photosynthetic apparatus, which fixes CO_2 and evolves O_2, is present in the vegetative cell but the oxygen-evolving component is absent from the heterocysts. Thus the heterocyst is a compartment from which the photo-generation of O_2 is excluded. Blue-green algae are normally aerobic (or microaerophilic; some can show highly developed anaerobic metabolism which will not concern us here). Therefore they generally need some oxygen for respiration and air can very likely diffuse into the heterocyst to some extent: presumably respiratory protection can cope with that.

    There exist filamentous blue-green algae such as *Plectonema* which can fix nitrogen but which do not possess heterocysts. Characteristically they are very sensitive to oxygen, and to illumination (which leads to high oxygenation as a result of photosynthesis), and only fix if the oxygen tension and illumination are low. *Anabena* thus represents a phototrophic microbe whose nitrogen fixation processes are better adapted to the aerobic way of life than those of *Plectonema*.

3.1.8   Nodules

The best-known nitrogen-fixing systems are the root nodules of leguminous plants, within which the bacteroid forms of *Rhizobium* fix nitrogen (see Chapter 5). Nodules which are good at fixing nitrogen are called 'effective' (the matter of effective and ineffective strains of *Rhizobium* is also discussed in Chapter 5) and effectiveness has been known since the 1940s to be associated with the presence of a red pigment in the nodules. This pigment, which renders nodules slightly pink and which can be seen very clearly when nodules are sliced in half, is a haemoprotein rather like the haemoglobin of mammalian blood. It is called 'leghaemoglobin'. Its importance remained an enigma until the mid 1970s when workers in Australia established its function. It proves to transport oxygen, just like haemoglobin, but its affinity for oxygen is so high that it delivers it to the rhizobia at a concentration which is harmless to their nitrogenase. Thus leghaemoglobin represents a sort of oxygen buffer: a molecule which, by combining reversibly with oxygen, prevents the accumulation of high concentrations of free oxygen while at the same time providing the oxygen necessary for metabolism of *Rhizobium*. The legume root nodule, then, is a highly specialized compartment in which nitrogen fixation and oxygen consumption are rendered physiologically compatible.

3.1.9   Other oxygen-restricting processes

The blue-green alga *Gloecapsa* is coccoid. It does not form filaments nor does it possess heterocysts. Yet it is surprisingly oxygen-tolerant. Under the electron microscope it shows a complex intracellular membrane structure, which may imply sub-cellular compartmentation, but some scientists have speculated that it separates photosynthesis in time: that it fixes nitrogen when-young, when its photosynthesis is still developing towards maximum O_2 production, and ceases fixation in a later period of maximal photosynthesis. The pelagic marine blue-green alga *Trichodesmium* forms mobile trichomes (rod-like microbes) which can aggregate in bundles if not too disturbed by the sea. Within these bundles, the central cells seem to lose photosynthetic activity and gain nitrogen-fixing activity so the cluster as a whole becomes a nitrogen-fixing microcosm in which photosynthesis takes place on the outside and nitrogen is fixed inside. This idea cannot be considered established until populations of *Trichodesmium* have been studied which are free of contaminant bacteria, for these might be fixing the nitrogen. Such populations have not yet been obtained.

3.2 The biological reductant

It has been fortunate for the purposes of biochemistry that nitrogenase interacts with the chemical sodium dithionite (see Chapter 2) and then performs its natural functions. Sodium dithionite is not present in living cells. The function of sodium dithionite, that of providing reducing power for the enzyme, is performed in living cells by specialized proteins called ferredoxins or flavodoxins. These are relatively small molecules of molecular weights ranging from 6000 to 24 000 which have in common the property of existing in oxidized and reduced forms which can be converted into each other easily. The reduced forms are strong reducing agents, capable of reducing many other biological molecules (given an appropriate enzyme to catalyse the reaction). The reduced forms usually react rapidly with air, becoming oxidized. Both types of protein can, in fact, exist in more than two oxidation and reduction states, but only two of these are important physiologically.

The ferredoxins are involved in a variety of biological processes, including photosynthesis in plants and pyruvate metabolism in anaerobic bacteria. Those concerned in nitrogen fixation belong to the 'four iron' class, which means that they have one, or more often two, clusters of four iron and four sulphur atoms in the molecule and it is a curious fact that the many iron atoms of nitrogenase (see section 2.1, Table 2) are in the form of similar iron-sulphur clusters. Changes in the valency of the iron atoms in the clusters are responsible for the special oxidation-reduction properties of the ferredoxins. It is not possible to discover precisely which iron atoms are involved because the whole cluster behaves like an oxido-reducible unit, but the kind of spectroscopy called electron paramagnetic resonance (see section 2.3) indicates that the ferredoxins of aerobic nitrogen-fixing bacteria such as *Azotobacter* behave slightly differently from those of anaerobes such as *Clostridium pasteurianum*.

In *C. pasteurianum*, ferredoxin is the actual protein which reacts with nitrogenase and provides the reducing power for the conversion of N_2 to NH_3. In *Azotobacter chroococcum* the situation is more complex. A ferredoxin is present, but a flavodoxin is the primary reductant for nitrogenase. This is a yellow protein which, when half-reduced, is a striking blue colour (the 'semiquinone' form). When fully reduced it is colourless. It is the colourless form which reacts with nitrogenase, providing the necessary reducing power, and the semiquinone form is produced. In the living cell, the flavodoxin probably cycles between the reduced ('quinol') and semiquinone form. The semiquinone form of *Azotobacter* flavodoxin is unusually stable to oxidation by air, which may be why this protein rather than a ferredoxin is particularly suitable to *Azotobacter*'s aerobic way of life.

Flavodoxins do not contain iron atoms; their oxido-reducible centre is a yellow, fluorescent molecule called a flavin.

3.3 The need for ATP

Nitrogenase requires biological energy as ATP in order to function, as was described in section 2.2. The source of this ATP is the normal metabolism of the microbe: catabolism of food leads to formation of ATP from ADP by a variety of chemical mechanisms, many of which are broadly understood but outside the scope of this chapter. If the ATP were not used for nitrogen fixation, it would be used for growth and multiplication. It follows that cultures of microbes which are fixing nitrogen are less efficient at converting food into cell material than are cultures which are using fixed nitrogen.

It is possible to study the efficiency with which microbes convert their food into cell material simply by growing them in an environment in which the extent of their growth is limited by the amount of food supplied. For example, a culture of *Klebsiella aerogenes* grown in optimal conditions at 37°C will convert 1 g of glucose into about 0.5 g (dried weight) of bacteria. Optimal conditions imply good aeration as well as growth in a continuous culture apparatus rather than the conventional laboratory batch culture; they also imply that the supply of glucose alone should limit the population density in the culture, not the supply of other nutrients such as phosphates, ammonium ions and potassium ions. With glucose, particularly when it is used by an anaerobic microbe, it is possible to know how many molecules of ATP are generated for each molecule of glucose consumed and hence, quite easily, one can know the number of ATP molecules consumed to yield a gram (dried weight) of bacteria. Details of the experiments and calculations are not necessary here; the important principle is that yield studies on continuous cultures of microbes limited by a single carbon source can give information on the efficiency with which ATP generated with the aid of that source is utilized for multiplication. If one compares carbon-limited populations of nitrogen-fixing bacteria which are using ammonium ions as their nitrogen source with similar populations which are fixing nitrogen, one finds that the yields with the latter are lower—sometimes substantially lower—and it is a reasonable assumption that most of this difference arises because ATP is diverted from biomass production to nitrogenase function.

Though such experiments are subject to appreciable error in practice, they give some interesting results. With *Clostridium pasteurianum* the yield difference corresponds to about 20 molecules of ATP diverted to reducing one N_2 molecule to two NH_3 molecules (the cell-free enzyme uses 12 to 15 ATP molecules: see section 2.2). With *Klebsiella pneumoniae*, some 30 ATP molecules are diverted to fixation. I described in section 3.1 how nitrogen-fixing *K. pneumoniae* can be grown with air provided its respiration is adequate to keep the dissolved oxygen near to zero. In those circumstances the yield increases, but this seems to be because the microbe can generate more ATP per molecule of glucose when provided with a little air; it does not become more efficient in its ATP economy for nitrogen fixation.

It is less easy to do comparable experiments with the aerobe *Azotobacter* for three reasons. One is general: aerobes are very much more efficient at generating ATP so the number of ATP molecules produced per gram of cells is difficult to calculate. The second arises because *Azotobacter* can alter the efficiency with which it generates ATP in response to the oxygen tension (see section 3.1.4). The third, and more drastic, is that respiratory protection leads to consumption of substrate connected neither with biosynthesis nor with nitrogenase function. Despite these difficulties it is possible to make approximations and extrapolations which provide an estimate of the minimum efficiency of *Azotobacter*, and the surprising result is that this organism has the most efficient ATP economy of all. Something in the region of 6 ATP molecules become diverted into nitrogenase function: the living organisms appear to be twice as economical as the cell-free enzyme and over three times as economical as the anaerobic nitrogen-fixing bacteria. Azotobacters not only have the most effective oxygen-excluding processes (sections 3.1.4 and 3.1.7), they also seem to be exceptional in their ATP economy. They seem to be among the most highly developed and best adjusted, free-living, nitrogen-fixing micro-organisms for the terrestrial environment as it exists today.

3.4 The need to control hydrogen evolution

Active cell-free nitrogenase evolves hydrogen from water if it has no nitrogen or analogous substrate to reduce (section 2.2). In fact, some hydrogen (though much less) is evolved even if nitrogen is present. The question then arises whether living nitrogen-fixing bacteria can prevent this side-reaction. At present the answer seems to be that many cannot: *Klebsiella* and *Clostridium pasteurianum* evolve hydrogen naturally during anaerobic growth and, if they are simultaneously fixing nitrogen, a proportion of the hydrogen comes by way of nitrogenase. The side reaction yielding H_2 may account for the low ATP efficiency of nitrogen-fixing cells of these organisms discussed in the previous section, since the ATP used for this process is effectively wasted. The enzyme responsible for normal hydrogen evolution (i.e. that formed in the normal metabolism of these bacteria, not via nitrogenase) is called hydrogenase, and it can catalyse both the uptake and evolution of hydrogen. *Azotobacter* evolves no hydrogen in normal circumstances and it contains a kind of hydrogenase which is not reversible: it only catalyses the uptake of hydrogen. This enzyme can be selectively inhibited by certain chemicals and, when this is done, *Azotobacter* cells start evolving hydrogen from their nitrogenase.

This is illustrated in Table 3; it is the sort of evidence which has led to the view that, though nitrogenase does evolve some hydrogen in the living cell, hydrogen is trapped by hydrogenase in *Azotobacter* and recycled. It can be used to generate ATP and/or reducing power. The use of recycled hydrogen to generate some extra ATP may contribute to the unusually efficient ATP economy of *Azotobacter* mentioned earlier.

Table 3 Acetylene and carbon monoxide cause *Azotobacter* to evolve hydrogen

Cultures of nitrogen-fixing *A. chroococcum* were incubated at 30°C in closed flasks with a large air space to which the gases shown were added. Hydrogen was measured in the air space by gas chromatography.

Added gas	nmol of H_2 evolved/hour
None	None
10% acetylene (C_2H_2)	20
5% carbon monoxide (CO)	440
10% C_2H_2 + 5% CO	760

Blue-green algae and photosynthetic bacteria show photo-evolution of hydrogen which is partially attributable to nitrogenase. A fascinating recent development has been the discovery that symbiotic nodule bacteria show a similar side reaction. In some plant-bacteria associations as much as 50% of their available ATP and reducing power can be lost in H_2 evolution (Table 4). In terms of the productivity of crops such as soya beans, this is a finding of great agricultural importance: a 'tight' (non-H_2-evolving) symbiosis obviously fixes more nitrogen per unit of solar energy than a 'loose' (H_2-evolving) one. Both the species of plant and the strain

Table 4 Formation of hydrogen by symbiotic nitrogen-fixing associations

Nodules were cut from the roots of leguminous nitrogen-fixing plants (see Chapter 5) and the amount of hydrogen formed measured in relation to their nitrogen-fixing activity.

Plant	% nitrogenase activity diverted to hydrogen evolution
Soya bean	55
Garden pea	39
Cowpea	< 1

of symbiotic bacterium can affect the tightness of the association in this respect. It is interesting that the non-leguminous associations, as far as they have been examined, are tighter than most of the leguminous associations.

3.5   The need to regulate nitrogenase synthesis

Nitrogenase is an expensive enzyme in terms of biological energy: it consumes a lot of ATP even in the relatively efficient *Azotobacter*. Therefore, nitrogen-fixing bacteria have developed means of regulating both its activity and its synthesis.

Relatively little is known about the details of the regulation of the activity of nitrogenase, but in outline the position seems simple. ADP, the product of utilization of ATP by nitrogenase, inhibits the enzyme. ATP is essential for the enzyme to work at all, so the ratio of ATP to ADP in the neighbourhood of the enzyme can determine the rate at which the enzyme works. In a test tube this 'control' process works convincingly; in the living cell the situation is very probably more complex—if only because the ratio of adenine nucleotide concentrations (ATP, ADP and AMP) has a controlling effect on the activity of many enzymes other than

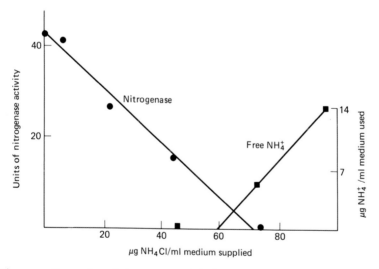

Fig. 3–4   Repression of nitrogenase activity by ammonia. *Azotobacter chroococcum* was grown in liquid media with increasing amounts of ammonium chloride (NH₄Cl). The nitrogenase activity (tested with acetylene) of the populations so obtained is given in arbitrary units. The amount of ammonium ion remaining in the medium after removal of the grown bacteria is also plotted: none was detected unless nitrogenase activity was completely repressed.

nitrogenase, and separate controls of the latter simply by the ratio of two of these nucleotides would thus be difficult. There is some evidence that carbamyl phosphate, an energy-rich substance formed from ATP in cells, has a regulatory function on the nitrogenase of *Clostridium pasteurianum*.

Regulation of the synthesis of nitrogenase is rather better understood at the molecular level, but detailed discussion will be postponed to section 6.3. At a descriptive level, ammonia (the primary product of nitrogenase activity) suppresses synthesis of the enzyme: if the environment contains sufficient ammonia for the microbe's needs, it does not make any nitrogenase. If there is less than sufficient, but still some, the microbe makes only sufficient nitrogenase to satisfy its requirement for fixed nitrogen. The situation is illustrated in Fig. 3–4. Other nitrogen sources—nitrates, urea, amino acids—often suppress nitrogenase synthesis equally well, probably because they are converted transiently to ammonia during metabolism of the cell. By growing organisms with minimal ammonia and letting them use it up, one can study rates of nitrogenase synthesis and learn that, in *Azotobacter* and *Klebsiella*, synthesis of both nitrogenase proteins is regulated together. When ammonia is absent, a 'message' is made in the cells telling its biosynthesis machinery to make the enzyme; this 'message' has a half life of about 4 minutes.

Azotobacter is good at protecting its enzyme from oxygen, but *Klebsiella* is less effective in this way. It would be biologically improvident for *Klebsiella* to make nitrogenase in air, only to have it destroyed by oxygen, and indeed it does not do so. As was mentioned in section 3.1.2, oxygen can repress nitrogenase synthesis by *Klebsiella* in a rather similar way to ammonia: neither protein is formed at all when detectable free oxygen is present in the environment.

4 The Free-living Microbes

The advent of the acetylene test provided microbiologists with a rapid and generally very effective tool for the assessment of the nitrogen-fixing potential of organisms, symbiotic systems and environments. New organisms and associations were discovered and, inevitably, some older ones came under suspicion and, in some cases, were rejected. The process of revision is continuing in the 1970s. In no case, yet, has a positive acetylene test given a false result *when applied to a living system* (many inanimate systems, such as chemical mixtures, will reduce acetylene to ethylene in water; examples are colloidal palladium under H_2, or a mixture of cysteine, sodium molybdate and sodium borohydride). The converse proposition, that all biological nitrogen-fixing systems give a positive acetylene test is so nearly true as to be reliable, but one exception (the methane-oxidizing bacteria discussed in section 4.3) is known and, as indicated in Chapter 3, adverse environmental conditions (e.g. high oxygenation) can obscure potential nitrogen-fixing ability in many microbes.

This chapter will be concerned with the organisms which fix nitrogen alone and which do not normally, or necessarily, have a symbiotic partner such as a plant. Table 5 gives a list of currently accepted organisms in this group; several will reappear in the next chapter because they can in fact form symbiotic associations; indeed, some are best known in a symbiotic context. The list includes no higher organisms at all: all are bacteria or blue-green algae and even very simple nucleated organisms such as yeasts or fungi are not represented (before about 1970, some yeasts and related microbes were thought capable of fixing nitrogen). It is important to be clear that free-living nitrogen-fixing microbes, at least when in the free-living state, are rarely of serious importance in the terrestrial nitrogen economy. Russian workers have estimated that an anaerobe such as *Clostridium pasteurianum* contributes perhaps 0.3 kg N/hectare/year to a soil, compared with 1000 times that amount contributed by a good leguminous association. *Azotobacter*, probably because of its need to divert subtrate into respiratory protection (section 3.1.4), is about as effective as *Clostridium pasteurianum* and the only free-living organisms of serious agronomic importance are the blue-green algae which, being capable of photosynthesis, are not limited by availability of carbon substrates in soil (these matters are discussed in more detail in section 4.4). Despite their generally rather low agronomic significance, the free-living nitrogen-fixing bacteria are of exceptional scientific importance and, since the majority of them can form productive associations of one kind or

Table 5   List of genera of microbes which include nitrogen-fixing species or strains

	Genus or type	Species (examples only)
STRICT ANAEROBES	Clostridium	C. pasteurianum*, C. butyricum
	Desulfovibrio	D. vulgaris, D. desulfuricans
	Desulfotomaculum	D. ruminis*
	(Methane-producing bacteria)	see text
FACULTATIVE (aerobic when not fixing nitrogen)	Klebsiella	K. pneumoniae, K. oxytoca,
	Bacillus	B. polymyxa, B. macerans
	Enterobacter	E. agglomerans
	Citrobacter	C. freundii
	Escherichia	E. intermedia
	Propionibacterium	P. shermanii, P. petersonii
MICROAEROPHILES (normal aerobes when not fixing nitrogen)	'Mycobacterium'	M. flavum*, M. roseo-album*
	Thiobacillus	T. ferro-oxidans
	Spirillum	S. lipoferum*
	Aquaspirillum	A. perigrinum*, A. fascicilus*
	Methanosinus	
	'Cowpea rhizobia'	
AEROBES	Azotobacter	A. chroococcum*, A. vinelandii*
	Azotococcus	A. agilis*
	Azomonas	A. macrocytogenes*
	Beijerinckia	B. indica*, B. fluminis*
	Derxia	D. gummosa*
PHOTOTROPHS (anaerobes)	Chromatium	C. vinosum
	Chlorobium	C. limicola
	Thiopedia	
	Ectothiospira	E. shapovnikovii
PHOTOTROPHS (facultative)	Rhodospirillum	R. rubrum
	Rhodopseudomonas	R. palustris
PHOTOTROPHS (microaerophiles)	Plectonema	P. boryanum
	Lyngbya	L. aestuarii
	Oscillatoria	
	Spirulina	
PHOTOTROPHS (aerobic)	Anabena	A. cylindrica, A. inaequalis
	Nostoc	N. muscorum
	Calothrix	
	(7 other genera of heterocystous blue-green algae)	
	Gloeocapsa	G. alpicola

* Signifies that all reported strains of the species fix nitrogen. For commentary, see text.

another, knowledge of their general microbiology is very valuable.

As well as including no higher organisms, Table 5 also includes no thermophilic microbes. These are specialized bacteria, which grow at temperatures well above 44 to 50°C, which is the upper limit for normal 'mesophilic' microbes. They are found in hot springs, fermenting compost heaps, geothermal waters, hot water systems and (surprisingly) ordinary soil; often they grow only at elevated temperatures. No thermophilic nitrogen-fixing bacteria are known, and it seems possible that there is an upper limit to nitrogen fixation of about 40°C.

4.1   The anaerobic bacteria

The first nitrogen-fixing microbe to be discovered was *Clostridium pasteurianum*, obtained by a famous microbiologist called S. Winogradsky in Paris in 1893. He called it *C. pastorianum*, meaning a spindle-shaped organism (*clostridium*) from the field, but its species name became transformed to *C. pasteurianum* (meaning pertaining to another micro-biologist, Louis Pasteur) in the 1920s. It metabolizes carbon sources such as glucose to butyric acid, CO_2, H_2 and minor products and is related to other butyrate-forming clostridia such as *C. butyricum* and *C. aceto-butyricum*. While by no means universal, ability to fix nitrogen seems not uncommon among the clostridia and they are widespread in soils, in decaying vegetable matter (e.g. compost heaps) and even in the rumens of ruminant animals such as cows or sheep (they do not fix nitrogen there, and are probably only itinerants, ingested with the food). They are obligate anaerobes; though sometimes tolerant of traces of oxygen, their metabolism can make no use of oxygen. In adverse conditions they form spores which resist heating and drying and these, no doubt, account for the ease with which *C. pasteurianum* can be isolated from almost any soil or water sample.

A more exacting group of anaerobes which includes nitrogen-fixing bacteria is the sulphate-reducing bacteria. These are organisms which use the oxygen atoms of sulphate ($SO_4{}^{2-}$) for respiration, producing sulphide and water. A typical overall metabolic reaction is:

$$2CH_3.CHOH.CO_2Na + CaSO_4 \longrightarrow$$

sodium lactate   calcium sulphate

$$2CH_3CO_2Na + CaS + 2H_2O + 2CO_2$$

sodium acetate   calcium sulphide   (8)

In most conditions the sulphide hydrolyses to the gas hydrogen sulphide, contributing to the repellent smell of highly polluted environments. They fall into two well-known genera, *Desulfovibrio* and *Desulfotomaculum* (recently a proposal for a third group, *Desulfomonas* has been made). *Desulfovibrio* includes several species within which are strains able to fix nitrogen, though the property is not universal within any species; within *Desulfotomaculum* only the species *D. ruminis* fixes nitrogen (but only two representatives of this species have in fact been examined and the other principal species (*D. nigrificans*) is entirely thermophilic).

Though most free-living nitrogen-fixing bacteria are not of great environmental importance, the sulphate-reducing genus *Desulfovibrio* does have a special ecological importance in the sea. This is probably the only indigenous marine nitrogen-fixing genus which is not phototrophic (i.e. does not require light) and it is considered to be the principal contributor of biologically fixed nitrogen to marine sediments. Other nitrogen-fixing bacteria, including *Desulfotomaculum*, butyric clostridia, klebsiellae and even azotobacters, can sometimes be found in off-shore waters, but they are washed into the sea with land water and are not truly marine nitrogen fixers.

Even more exacting anaerobes than the sulphate-reducing bacteria are the methane-producing bacteria. These are plentiful in fermenting sewage sludge and also in the bottom muds of ponds and lakes, particularly if these are subject to periodic pollution by, for example, leaf fall. They are responsible for the formation of marsh gas in ponds and marshes. They must be mentioned because, though they were long believed to be able to fix nitrogen, the one strain on which this belief was based proved to be a mixed population. However, whichever partner was responsible for fixation was a very strict anaerobe.

Certain photosynthetic anaerobes fix nitrogen and are mentioned in section 4.4.

4.2   The facultative bacteria

These are a physiological group of bacteria which are able to grow either with or without the oxygen of air when provided with fixed nitrogen, but which can only fix nitrogen anaerobically. For this reason it took many years before their ability to fix nitrogen became firmly established. There is now no shadow of doubt and, indeed, it seems that they are very common nitrogen fixers in soils and waters.

A member of this group, *Klebsiella pneumoniae*, has already appeared frequently in earlier chapters. Some microbiologists classify the strain discussed as *K. oxytoca*. Numerous strains and species of *Klebsiella* have been examined for nitrogen fixation and it is a curious fact that the property is distributed over about 50% of them with no particular correlation with the source of the strain: a *Klebsiella* from the gut is as likely—or as unlikely—to fix nitrogen as one from a soil or water sample.

Klebsiella belongs to a group of bacteria called 'coliforms', so-called because they resemble in appearance a bacterium called *Escherichia coli* normally present in the guts of animals, including man. No fully authenticated natural strains of nitrogen-fixing *E. coli* have been isolated, though laboratory strains able to fix nitrogen have been constructed genetically (see Chapter 6). Apparently the species *E. intermedia* can fix nitrogen. Among other coliforms, the genus *Citrobacter* (found in soils, foods and the intestines of termites) includes nitrogen-fixing strains and so does the genus *Enterobacter* (which also inhabits intestines, as well as soils and waters). Some coliform plant commensals such as the species *Erwinia herbicola*, include nitrogen-fixing strains but the property is not universal.

The genus *Bacillus* is a second major group of facultative bacteria which includes nitrogen-fixing strains. *Bacillus polymyxa* and *Bacillus macerans* are two species among which nitrogen fixation seems common (but again not entirely universal) and reports of fixation by other species of *Bacillus* exist. The bacilli are characterized by their ability to form spores and thus survive well in environments subject to desiccation or periodic heating. Their behaviour has something in common with clostridia and, indeed, the borderline between air-sensitive bacilli and aerotolerant clostridia is difficult to define. The propionic bacteria mentioned in Table 5 are reported by Russian workers to fix nitrogen anaerobically. They occur in fermenting milk and their possession of this property is rather surprising.

Certain phototrophic bacteria are facultative anaerobes and only fix nitrogen anaerobically; they are discussed in section 4.4.

4.3   The aerobes

The best-known aerobic nitrogen-fixing bacteria are members of the genus *Azotobacter*; indeed, *A. chroococcum* (Figs. 4–1 and 3–2) was the second free-living nitrogen-fixing microbe to be discovered after Winogradsky's *C. pasteurianum*. It was reported by the famous Dutch microbiologist M. W. Beijerinck in 1901. Today the '*Azotobacter* group' comprises several genera: *Azotobacter*, *Azomonas*, *Azotococcus*, *Beijerinckia* and *Derxia*; they are all rather similar in appearance and physiological character and none is capable of growth without air. They also look rather alike: large granular microbes, ovoid or rod-shaped, sometimes motile, which form spreading, glutinous colonies on nitrogen-free agar media. In general, the azotobacters are well adjusted to oxygen: they fix nitrogen in ordinary air relatively efficiently and the only representatives of the group which do not fix nitrogen are mutants artificially prepared in the laboratory. Some, such as *Derxia*, have difficulty in initiating growth in air (see section 3.1.3.) and within the group there is in fact a range of oxygen-sensitivities which may reflect the extent of their adjustment to an aerobic way of life. If *Derxia* is most sensitive to oxygen stress, then *Beijerinckia* is intermediate and *A. chroococcum* and *A. vinelandii* are least oxygen-sensitive. Even these, however, can be inhibited by really high oxygenation.

Most of the aerobic nitrogen-fixing bacteria are less well-adapted to air

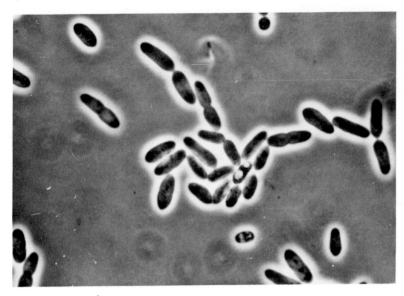

Fig. 4–1 Photomicrograph of *Azotobacter chroococcum*. Each microbe is about 6 μm long. (Courtesy of Dr P. J. Dart and Dr Johanna Döbereiner.)

than the azotobacters and were recognized only relatively recently. The reason is that, when they fix nitrogen, they behave like microaerophiles. The microaerophilic character of *Mycobacterium flavum* (named for the yellow colour of its colonies) was mentioned in section 3.1.3 and two or three 'mycobacteria' of rather similar properties were described along with it when it was discovered by scientists in the U.S.S.R. in 1961. Mycobacteria are relatives of the causative organisms of tuberculosis and leprosy, but it is highly probable that *M. flavum* and its relatives have been mis-classified and have no relationship to authentic mycobacteria. *M. flavum* certainly has a lot in common with a recently-discovered group of aerobic nitrogen-fixing bacteria called *Corynebacterium autotrophicum*, many strains of which can grow chemotrophically (see section 1.5) with gaseous hydrogen as an energy source, deriving energy from its oxidation to water. They, too, are very sensitive to oxygen.

Another broad group of aerobes which behave like microaerophiles when fixing nitrogen is the spirilla, represented by the species of *Aquaspirillum* listed in Table 5 and *Spirillum lipoferum*, which is involved in symbiotic associations with grasses (see section 5.4). They are generally rather large, curved or corkscrew-like bacteria which often grow as a pellicle some way below the surface of a culture medium (where, presumably, the dissolved oxygen tension is just right). Reports of nitrogen-fixing spirilla had been in the scientific literature for many years but, until recently, their oxygen sensitivity made their ability to fix nitrogen difficult to establish with certainty. Similar considerations apply to *Thiobacillus ferro-oxidans*, a remarkable microbe which grows in acid mine waters (pH 2 to 4) and obtains energy by oxidizing dissolved ferrous ions to ferric ions. Like the nitrifying bacteria (section 1.5), this organism is a chemotroph: the energy of iron ion oxidation enables it to convert CO_2 to organic matter and to multiply. Strains of this resourceful microbe can fix nitrogen, but only at low dissolved oxygen tensions.

The methane-oxidizing bacteria were also long suspected of being able to fix nitrogen—partly because the soil around blow-holes of natural gas (methane) is generally more fertile and richer in fixed N than elsewhere. Many strains can, indeed, fix nitrogen and they show great sensitivity to oxygen, but authentication of these was delayed for a different reason. They provide, in fact, the only example in which the celebrated acetylene test for nitrogen fixation breaks down. Acetylene actually inhibits the enzyme system responsible for methane oxidation so, if these bacteria are grown with methane and tested with acetylene, the result is negative: they cannot generate any ATP from methane and so cannot use their nitrogenase even if they have it. All strains can use methanol (methyl alcohol) in place of methane so, if this substrate is provided, their possession of nitrogenase shows itself clearly.

Perhaps the most dramatic example of microaerobic fixation occurs among the rhizobia, the symbiotic partners of leguminous plants. For many decades a dogma existed to the effect that rhizobia never fix nitrogen away from a host plant. In the mid 1970s this view was overthrown and scientists in Australia and Canada successfully grew certain slow-growing types of rhizobia such as 'cowpea' strains (see section 5.1 for the classes of rhizobia) in media entirely free of plant material. The secret is that these microbes are very bad at respiratory protection (see section 3.1.4) and therefore their nitrogen fixation is highly oxygen-sensitive (they grow in air without difficulty given fixed N as glutamate or nitrate, of course). Only with the most careful control of the dissolved oxygen concentration will they fix nitrogen in liquid media; on solid (jellified) media things are a little easier provided some fixed nitrogen is around, because colonies of bacteria grow and within these at least some of the microbes find the oxygen tension just right for fixing nitrogen. Though there is now no doubt that these bacteria can make nitrogenase and fix nitrogen outside a host plant, at the time of writing (late 1976) they had never been grown in media entirely free of fixed nitrogen. They need a small amount of ammonia, glutamine or some other fixed nitrogen source to start them off; whether this is a subtle reflection of their extreme oxygen sensitivity or a special physiological requirement for fixation is not yet known.

4.4   The phototrophs

As explained in section 1.5, phototrophs are autotrophs which make use of light to fix CO_2 and multiply. These are phototrophic representatives of all the classes of microbes discussed in the preceding section of this chapter. The strict anaerobes such as *Chromatium* and *Chlorobium* are sulphur bacteria: they oxidize sulphides, elemental sulphur or thiosulphates to sulphates while absorbing light with chlorophyll and carotenoid pigments (like plants). The energy therefrom is used for growth and can be coupled to nitrogen fixation. They are coloured: *Chlorobium* species are green, *Chromatium* and *Thiopedia* are red or purple. Probably many other types of coloured sulphur bacteria can fix nitrogen; all are obligate anaerobes. The coloured non-sulphur bacteria, of which *Rhodospirillum rubrum* and *Rhodopseudomonas palustris* (both purple) are examples, grow phototrophically, but only anaerobically, and their photosynthetic processes do not lead to evolution of oxygen. Then they can fix nitrogen. It is possible to grow these organisms heterotrophically (i.e. with organic substrates such as sugar or alcohols) and aerobically, but only in the dark. In these circumstances they do not fix nitrogen.

Anaerobic and facultative phototrophic bacteria may be of minor importance to the terrestrial nitrogen cycle except in highly specialized environments such as sulphur or polluted waters, but by far the most important phototrophs from an ecological viewpoint are the blue-green algae (they are in fact true bacteria). They differ from the coloured sulphur and non-sulphur bacteria in being able to grow phototrophically in air (though some can actually mimic the sulphur bacteria and grow phototrophically and anaerobically with sulphide). They also have a more sophisticated photosynthetic equipment which, like that of higher plants, evolves oxygen as a product of photosynthesis. Thus they face the problem, discussed in section 3.1.7, of excluding photo-generated oxygen from nitrogenase, and they can be grouped according to the effectiveness with which they do this. As was described in section 3.1.7, the most sophisticated have compartments called heterocysts to which nitrogenase is restricted. Eight genera of heterocystous blue-green algae are known and, though nitrogen fixation has not been confirmed in all heterocystous blue-green algae, the majority fix nitrogen readily in air at normal illuminations. *Anabena cylindrica* has been studied in depth in several laboratories and its properties are probably typical. The heterocysts are formed only in nitrogen-deficient environments (see Fig. 3–3); they appear at regular intervals along the filament and the nitrogenase activity of the population is proportional to the heterocyst count. Nitrogenase has been extracted and partly purified from heterocysts; their lack of photosystem II (responsible for the photo-evolution of O_2: see Studies in Biology No. 37) was mentioned in section 3.1.7. The evidence for a role for heterocysts as oxygen-restricting compartments for nitrogenase is overwhelming.

An aberrant genus is *Gloeocapsa*. These are unicellular blue-green algae, little green spheres, with no heterocysts. Though sensitive to oxygen, they are not exceptionally so. Ability to fix nitrogen is rare within the group (though unequivocal when present) and, as discussed in section 3.1.9, these organisms may have an unusual oxygen-restricting process.

Several genera of non-heterocystous blue-green algae exist and, characteristically, they are microaerophilic when fixing nitrogen. They also dislike high illumination levels. They are frequently found in association with other microbes, in shaded or polluted waters, and they are clearly nitrogen-fixing phototrophs at a less sophisticated stage of evolution than the heterocystous blue-green algae.

As stated earlier, the blue-green algae are ecologically by far the most important free-living nitrogen-fixing bacteria. The relative unimportance of free-living nitrogen-fixing bacteria in the terrestrial nitrogen economy was emphasized at the opening of this chapter. For reasons which range from inefficient ATP production and/or utilization to massive respiratory protection, the ordinary nitrogen-fixing bacteria rarely receive sufficient energy-providing substrates in nature to upgrade the nitrogen status of the environment significantly. The blue-green algae are in a different position: they fix nitrogen by using solar energy and, provided they are not in the dark, energy ceases to be the critical limitation. Though their major ecological contribution to the nitrogen cycle is made symbiotically (see the next chapter), they probably contribute, as free-living organisms, some hundred times as much N to a soil as do bacteria such a *Clostridium pasteurianum* or *Azotobacter Chroococcum*. An estimate made by Russian workers of 30 kg N/hectare/year (cf. 0.3 kg quoted for *C. pasteurianum* at the beginning of this chapter) for the contribution of free-living blue-green algae to soils is probably an underestimate because it was made before fixation by the non-heterocystous species was discovered. Blue-green algae are the first to colonize arid, ice-bound and devastated areas when conditions begin to improve; they pioneer biological colonization of rocks, buildings, newly exposed sand, chalk and other infertile substrata; they often colonize living and decaying plant surfaces such as tree-trunks. They even fix nitrogen at the surface of ordinary soils, in puddles after rainy weather. They are probably the only important aerobic nitrogen-fixers in the open sea, as well as on rocks and coral reefs.

4.5   The 'ghosts'

To conclude this chapter it is necessary to say something about several groups of microbes which were once thought to fix nitrogen but which

are now known not to do so. Sometimes called 'ghosts' by
microbiologists, they include yeasts and some other simple fungi, species
of the genera *Pseudomonas* and *Azotomonas* and several species less clearly
identified. They all appear to grow successfully on (or in) nitrogen-free
media, and they often form thick, glutinous colonies like those of an
authentic nitrogen fixer such as *Beijerinckia*. Fig. 4–2 is a close-up

Fig. 4–2   Colonies of a 'ghost' on nitrogen-free agar. These colonies bear a
superficial resemblance to nitrogen-fixing bacteria (see Fig. 3–2) but consist
largely of mucoid material and very few microbes. (Courtesy of Professor H. J.
Evans.)

photograph of such a colony. Often the N-content of the environment is
augmented after growth of such 'ghosts'. They appear for two principal
reasons:

(1) Some yeasts and pseudonomonas are exceptionally good at utilizing
traces of fixed nitrogen present as contamination in bacteriological
reagents or in ordinary air. Town air contains oxides of nitrogen
and free ammonia; a colony on a Petri dish culture can scavenge
these gases, converting the nitrogen oxides to nitrate or nitrite,
assimilating the ammonia, thus facilitating further diffusion of
such gases towards itself. When this happens, the bacteria are seen
under the microscope to be very sparse in the colony—most of it is
mucilaginous carbohydrate. The N-content of an ordinary
microbe is 12 to 15% of its dry weight, but such 'ghost' bacteria can

have as little as 3 to 6% N. In one case known to the writer, a yeast 'ghost' had only 1% N and gave a most convincing impersonation of a nitrogen-fixing organism.

(2) A happily less frequent reason for a 'ghost' is contamination. *Clostridium pasteurianum* cannot grow in air but it can multiply inside colonies of some other bacteria such as *Pseudomonas*. Instances are known in which putative nitrogen-fixing aerobes were actually non-fixers harbouring undetected nitrogen-fixing anaerobes within their colonies. Such associations are misleading to the scientist, but they may reflect an important ecological situation if the two types of microbe assist each other. But this topic is more appropriate to the next chapter.

5 The Plant Associations

The existence of symbiotic nitrogen-fixing systems, in which bacteria fix nitrogen in association with the roots of plants, was suspected by Sir Humphry Davy early in the nineteenth century, but was only confirmed in 1886–8 by the German scientists H. Hellreigel and H. Wilfarth. They are now known to be much the most important contributors to the nitrogen cycle, from both ecological and agricultural points of view. The associations formed by leguminous plants in which soil bacteria of the genus *Rhizobium* colonize nodules on their roots (Fig. 5–1) are the best known and, since they include clover, lucerne (alfalfa), pulses (peas and beans) and so on, they are traditionally the most important to agriculture. In recent years, however, new types of association have been discovered and some known ones have become recognized as being of unsuspected importance. A few have proved not to be real nitrogen-fixing symbioses at all.

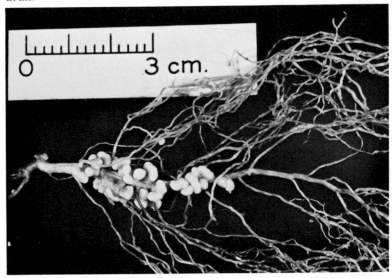

Fig. 5–1   Root nodules of the purple vetch (*Vicia atropurea*). (Courtesy of Dr P. J. Dart.)

5.1   The legumes

The family Leguminosae is a group of flowering plants found in both temperate and tropical zones. Botanists believe that they were originally tropical. They range from small but widespread plants such as clover, through flowering plants such as lupins and bushes such as brooms, to shrubs and trees (e.g. *Acacia*). The botanical properties of the Leguminosae will not be discussed here since they are available in any advanced botany textbook. The family has been divided into three major sub-families, the Papilionaceae (comprising most of the known species), the Mimosaceae and the Caesalpinaceae. Between 80 and 90% of the species in the Papilionaceae form nodules, but only about a quarter of the Mimosaceae and relatively few of the Caesalpinaceae. Well over 12 000 species of Leguminosae are known, the majority of which fix nitrogen; it is a sobering thought that less than 50 have been exploited for agricultural purposes and of these about 5 are regularly used in agriculture to up-grade soil. Table 6 lists a selection of familiar or agriculturally important members of the family Leguminosae.

Both leguminous plants and species of *Rhizobium* show a degree of specificity: certain types of plant are colonized by certain types of bacteria. A classification of rhizobia into 'cross inoculation' groups is given in Table 7, but it is not very rigid. The broad classification into fast-growing and slow-growing types seems to reflect real microbiological differences between the species, but the narrower classification has some degree of overlap within groups. Until recently, scientists believed that the Leguminosae were the sole hosts for *Rhizobium* but, in 1973, an instance

Table 6   Selected list of Leguminosae

Papilionaceae (temperate)	
Pisum sativum	Garden pea
Vicia faba	Broad beans
Phaseolus vulgaris	French or kidney bean
Trifolium repens	White clover
Medicago sativa	Lucerne
Ulex	Gorse
Lupinus polyphyllus	Garden lupin
Lotus corniculatus	Bird's foot trefoil
Melilotus officinalis	Melilot
(tropical and sub-tropical)	
Glycine max	Soya bean
Arachis hypogaea	Ground nut (peanut)
Cicer arietinum	Chick-pea
Mimosaceae	
Acacia	Includes the 'wattles'
Xilia dolabriformis	Ironwood of India
Mimosa pudica	Sensitive mimosa
Caesalpinaceae	
Cassis fistula	Senna
Tamarindus indica	Tamarind
Cercis siliquastrum	Judas tree

was discovered in which a cowpea *Rhizobium* nodulates *Trema aspera*, a non-leguminous tropical plant. *Trema* is at present the only non-leguminous genus known to be colonized by rhizobia.

The dogma that rhizobia never fix nitrogen except within the nodule of a plant was widely accepted for many decades until it was overthrown in the mid 1970s (see section 4.3). It crumbled in two stages. Tissue cultures of plant cells are not difficult to obtain and rhizobia were first shown to be

Table 7   Cross inoculation groups of *Rhizobium*

	Organism	Examples of host plants
FAST-GROWING TYPES		
Alfalfa group	*R. meliloti*	Melilot, lucerne
Clover group	*R. trifolii*	Clover
Pea group	*R. leguminosarum*	Peas, lentils, some beans (*Vicia*)
Phaseolus group	*R. phaseoli*	Beans (*Phaseolus*)
SLOW-GROWING TYPES		
Lupin group	*R. lupini*	Lupins
Soybean group	*R. japonicum*	Soya beans
Cowpea group	'Cowpea rhizobia'	Cowpea, peanuts, acacia, etc.

able to fix nitrogen in association with tissue cultures of a leguminous plant (soya bean). Then an active association with a tissue culture of a non-legume (tobacco plant) was demonstrated and, within months, conditions were discovered for fixation without plant material at all. As was described in section 4.3, slow-growing rhizobia prove to be highly sensitive microaerophiles when fixing nitrogen alone. Ability to fix nitrogen *ex planta*, as it is termed, is principally restricted to the slow-growing class of which the 'cowpea' group is typical; fast-growing rhizobia such as *R. meliloti* have not so far been induced to fix nitrogen away from a plant. Perhaps this is only a matter of time.

Despite the recent discovery of fixation *ex planta*, the really important process for biological productivity is conducted in symbiosis with a plant. Nearly all soils contain rhizobia, normally not fixing nitrogen, and these colonize appropriate plants at about the stage in the plant's growth season when leaves begin to appear. The plant produces a substance attractive to *Rhizobium* and the microbe, in its turn, produces a hormone-like substance which causes the root hairs to curl. At a later stage certain root hairs become infected and the bacteria invade the root, multiplying as they spread along the hair into the root tissue. Here they colonize special types of plant cells, around which the nodule tissue is ultimately formed. The cells become engorged with bacteria which then cease growing and often assume unusual shapes. In this state the bacteria are

then called 'bacteroids' and they are rich in nitrogenase. Fixation of
nitrogen then continues until (usually) seed formation, when it ceases
and the nodules become senescent. Some of the bacteroids are incapable
of further multiplication, but some remain viable and, as the nodule (or,
in annuals, the plant) ages and dies, these survive in soil for the next
season's growth.

In detail, the colonization process differs in character and timing from
plant to plant. There is some evidence that certain plant proteins called
lectins can interact with the specific rhizobia which colonize that plant and
that this property enables the plant to recognize and admit the correct
type of *Rhizobium*. A given plant, such as soya bean, can be infected by a
variety of strains of *Rhizobium*, and these can differ in their ability to fix
nitrogen efficiently. A really effective strain, as it is called, forms healthy-
looking nodules with a distinct pink colour; if the nodules are cut open,
the interior is distinctly red. Ineffective rhizobia form pallid, sometimes
greenish, nodules. Once a plant has been colonized by a certain strain of
Rhizobium, others seem to be excluded, so it is important in agriculture to
ensure that the rhizobia which infect a crop plant are effective. In many
areas the natural rhizobia are relatively ineffective, and cultures of good,
effective *Rhizobium* damp-dried in peat are now available to farmers for
'pelleting' seeds: for coating them with the live, effective rhizobia. Even
so, the local rhizobia can sometimes compete with the introduced ones
and supplant them. Means for supplying live, effective rhizobia to crop
plants can assume great agricultural importance, particularly in under-
developed countries, and really reliable means for doing this have still to
be developed.

One problem arises from the very effectiveness of the rhizobia
themselves. Leguminous plants do not admit rhizobia if the soil is
reasonably rich in fixed nitrogen. If a plantation of peas, for example, has
successfully nodulated on a poor soil and given a good harvest, much of
the non-harvestable material (roots, stems and leaves) will die, decay and
add fixed nitrogen to the soil, so up-grading it. In the next year,
nodulation will be discouraged by this available nitrogen, and the
effective rhizobia will tend to die out compared with the natural strains.
This 'second year death' can, in extreme cases, cause the soil to revert to
its low status by the third year. The practice of agriculture with symbiotic
systems is different from that with fertilizers, and requires different
outlooks on the part of the farmer. Other problems, such as the
responses of the bacteria to herbicides, drying, freezing and so on, need
consideration, too. On the other hand, cost of fertilizer (both material
and transport costs) becomes minimal and pollution problems resulting
from escape of fertilizer into wells, streams, rivers and so on are much
diminished. For a discussion of the relative advantages and disadvantages
of biologically versus chemically fixed nitrogen the reader should consult
more specialized works. He will discover that energy, transport and

pollution costs weigh heavily against chemical fertilizers but that they can
never be abandoned completely.

As was discussed in section 3.1.8, the importance of the red colour in
effective nodules is now understood. It is due to leghaemoglobin, a
protein which supplies oxygen to the bacteroids but avoids oxygen-
damage to their nitrogenase. The legume nodule is a complex biological
structure for protecting rhizobial nitrogenase from oxygen. It is
interesting that leghaemoglobin is produced by the plant, not the
bacteria. There is no shadow of doubt, however, that the nitrogenase
enzyme is produced by the bacteroids. The enzyme from *R. japonicum*
bacteroids has been extracted and substantially purified; like the
nitrogenases of free-living organisms discussed in Chapter 2, it consists of
two iron-containing proteins, one large and containing molybdenum in
addition. It is oxygen-sensitive and consumes ATP; it reduces the usual
substrates; it is so similar to the nitrogenases of free-living bacteria that its
components will cross-react with those of *Azotobacter*.

5.2   The nodulated non-legumes

The fact that a number of non-leguminous plants have root nodules
has been known for many decades. With some, such as the sub-tropical
conifer *Podocarpus*, the 'nodules' are associated with a symbiotic fungus
(mycorrhiza; see 5.5.2). Others may be gall-like products of microbial or
viral infection. But nearly 120 species are now known, principally trees
and shrubs, whose root nodules harbour nitrogen-fixing bacteria which
are not rhizobia. Table 8 lists some examples of such plants; the best
known example is *Alnus* (alder) and the symbiosis is sometimes called the
'alder type symbiosis'. They have a wide geographical distribution and
often colonize infertile or devastated areas. One of the first to be
recognized was *Eleagnus*, in the late nineteenth century, but knowledge of
the physiology of non-leguminous nitrogen fixation has progressed
relatively slowly—even the ecological importance of these plants, now
known to be considerable, has only been recognized in the past decade or
so. There are various reasons for the slow progress in understanding the
symbiosis: the plants are tough, woody and often slow-growing; many
inhabit unusual environments. But the principal difficulty has been the
symbiotic microbe: despite many efforts, the nitrogen-fixing microbe has
never been isolated in pure laboratory culture. It can be seen as a
filamentous growth in the nodule, and it seems to resemble a class of
bacteria called Actinomycetes. Crushed nodules can be used to infect
seedling plants, and evidence has been obtained that the microbe can
travel in soil from one plant to another but, though many actinomycetes
have been isolated from local soils and from active nodules, some
behaving very like the true symbiont in various ways, by 1976 none had
yet been used successfully to re-infect a plant.

Table 8   Some nitrogen-fixing, nodulated, non-leguminous plants

Genus	Common name of an example
Alnus	Alder
Casuarina	Australian pine
Ceanothus	Californian lilac
Eleagnus	(*E. maculata*)
Myrica	Bog myrtle
Hippophae	Sea buckthorn
Purshia	Bitter brush
Dryas	(*D. octopetala*)

Nevertheless, the organism has been given the name Frankia and some species have been described on the basis of their infectivity to plants. For, like rhizobia, the microbes show 'cross-inoculation' groups: for example, microbes which nodulate *Alnus* are said not to nodulate *Eleagnus* or *Ceanothus*, though they will infect *Myrica*. But *Myrica*'s organism does not infect *Alnus*. However, as with rhizobia, there is disagreement about the reliability of the information on which such beliefs are based and a systematic nomenclature and classification must await isolation of the microbes.

There is some evidence for restricted geographical distribution of the microbes. *Ceanothus* is regularly nodulated in the USA but rarely in Europe. *Alnus* is almost always nodulated in Europe but European *Myrica* by no means always has nodules.

The non-leguminous nitrogen-fixing systems are of modest agricultural importance today, partly because their potential value is not widely known. Yet their input of N on a global scale must be enormous: thousands of square kilometres of Washington State and Oregon, USA, are populated by *Alnus*, *Ceanothus* and *Purshia* together. *Alnus* alone can add 100 kg N/hectare/year to soil from leaf fall alone (about one third the input of a good fertilizing legume such as lucerne) so the combined contribution must be impressive. Plants in this group have little agricultural potential as crops or forage, but alder could be valuable in forestry. Scrub alders have been used traditionally to up-grade poor soils in Asia, and of the trees used to stabilize dykes in Holland, every third is customarily an alder. The exploitation of these symbioses obviously requires further attention.

With so elusive a symbiont, the biochemistry and physiology of the alder-type symbioses is obviously little studied. In recent years cell-free preparations of nitrogenase have been obtained successfully from *Alnus* nodules but they are difficult to handle and nothing is known of the enzyme beyond that it shows the familiar oxygen sensitivity. There is no evidence for leghaemoglobin in such nodules, nor for any similar protein, but it seems likely in principle that something performing a function analogous to leghaemoglobin must be present.

5.3   The algal symbioses

The ecological importance of blue-green algae as free-living nitrogen fixers was emphasized in section 4.4. Numerous instances are now known in which a blue-green alga forms an association with a plant. It is curious that, whereas the rhizobia and Frankia species form associations only with the most advanced plants, the dicotyledenous Angiosperms, the blue-green algae tend to favour the more primitive members of the plant kingdom.

5.3.1   The lichens

The most primitive plant-microbe associations of this kind are the lichens (pronounced 'lykens'), in which a nitrogen-fixing blue-green alga associates with a fungus. Since it is a little difficult to judge which is the host and which is the symbiont, the terms 'phycobiont' and 'mycobiont' are used to describe the algal and fungal partner respectively. In fact, most lichens do not fix nitrogen: the phycobiont is a true green alga. Rather less than 10% of known lichens have a blue-green alga as phycobiont (often *Nostoc* or *Calothrix*: see Table 5). As would be expected, light is necessary for nitrogen fixation by lichens, but the organisms otherwise tolerate extreme conditions. In antarctic species, nitrogen fixation takes place at 0°C and the organisms survive extended freezing; they also survive drying but do not fix nitrogen until they become damp again. They colonize the most unpromising environments: rocks, stones, walls and roofs of buildings and they are the principal vegetation of some arctic and tundra regions. As 'reindeer moss', large sub-arctic lichens are an important food source for reindeer.

5.3.2   Liverworts

The liverworts include species which have blue-green algae of the genus *Nostoc* associated with the underside of the thallus. There has been some argument about whether fixation of nitrogen by the *Nostoc* actually benefits the plant but recent work with the isotope of nitrogen $^{15}N_2$ seems to have confirmed modest incorporation of fixed nitrogen into the plant.

5.3.3   Pteridophytes

Azolla is a tiny water fern which grows in tropical waters, often to a high density. It is only 2 or 3 mm across and it harbours a blue-green alga of the genus *Anabena* (see Table 5). The microbe inhabits a cavity at the base of the leaf-like frond and is particularly rich in heterocysts (see section

3.1.7) which presumably protect its nitrogenase from the considerable oxygenation which its own and the plant's combined photosynthesis can develop in the tropical sunshine. The symbiotic *Anabena* is very difficult to grow away from its plant host, but this has been done successfully on occasions; the fern can grow quite well without the alga if fixed N is available. In nature, the association is very effective and *Azolla* is an important green manure for rice culture: in the high season, the rate of input of N into a rice paddy can equal that of a good legume crop on land (introduced into temperate zones, *Azolla* can still add substantial amounts of N to lake waters). It is one of the most promising systems for the improvement of tropical and sub-tropical agriculture.

5.3.4 Gymnosperms

A number of species of this phylum, among the family of Cycads, form nodule-like structures on the roots near, but beneath, the soil surface. One of the best described is a Cycad called *Macrozamia*; its roots develop club-shaped excrescences (called 'coralloid' roots, from their resemblance to coral) which include a deep green zone. There is no doubt that these roots are well-colonized by nitrogen-fixing blue-green algae such as *Nostoc* and that they fix nitrogen; the curious thing is that the process is not necessarily light-dependent. Glasshouse plants do fix more nitrogen if the roots are illuminated but field samples do not necessarily show such an effect.

5.3.5 Angiosperms

A single example is known of a higher plant which forms an algal association. *Gunnera* is a sub-tropical plant which is invaded by a species of *Nostoc*. As a result, nodules are found at the bases of the leaves and within these nitrogen is fixed and made available to the plant. As with *Azolla*, the *Nostoc* is particularly rich in heterocysts in the symbiotic state, possibly because its location is one to which photo-evolved oxygen has ready access.

5.4 Associative symbioses

This is a name for a rather vague group of recently discovered symbiotic systems in which there is some interdependence between the partners though both can grow satisfactorily apart. They involve grasses (Gramineae) principally, and the first one to become established scientifically was between a sand grass called *Paspalum notatum* and a species of *Azotobacter*, *A. paspali*. *Paspalum* grass is widespread and persistent in tropical and sub-tropical areas such as Brazil and Australia; it is of little value as a forage grass but some related grasses could be of value in animal production. Many cultivars of *Paspalum* form an association with *A. paspali* which is rather specific: *A. paspali* grows around the roots of the plant, though the dominant nitrogen-fixing microbes in the surrounding soil are usually different, *Beijerinckia* or *Derxia*, for example. In due course the azotobacters form a sheath, partly mucilaginous, over the roots and cease growing, though they do not cease fixing nitrogen. Those plants which are able to form this association show considerable benefits in vigour, dry weight and nitrogen content. It is interesting that the discoverer of the symbioses, Dr Johanna Döbereiner of Brazil, suspected its existence several years before she was able to confirm it with the acetylene test and with isotopic nitrogen. The reason is that the association is very sensitive to oxygen: if you dig up plants and test them as you might test legumes, with acetylene or isotopic nitrogen in air, the nitrogenase of *A. paspali* ceases functioning. Only if tests are performed with a low level of oxygen—probably closer to conditions down in the soil—are good positive results obtained. The input of N to soil by natural *Paspalum* associations is probably small, but the acetylene test indicates that some 90 kg N/hectare/year could potentially be brought to soil by this system—about 30% of a good leguminous crop.

A second well-established associative symbiosis occurs with another tropical/sub-tropical grass called *Digitaria decumbens* and was reported first in 1974. In this case the microbial partner is *Spirillum lipoferum* (see Table 5), a bacterium which is widespread in both temperate and tropical soils. The spirilla do not form a sheath around the roots of *Digitaria* but in fact they invade the root tissue, forming a layer beneath the epidermis where they cease multiplying but continue to fix nitrogen. There is no doubt that they supply nitrogen to the plant, but the scale on which they do this is not clear at the time of writing and may normally be small. The *Spirillum–Digitaria* association caused considerable excitement in 1975 because of reports that the microbe would associate with some varieties of maize. The prospect of nitrogen-fixing cereal crops was thereby opened, which, if real, would be of enormous benefit to world food production. More sober reports indicate that the bacteria are indeed catholic in their tastes: they will associate with some cultivars of maize, sorghum or millet; however, their effect on the N-content of the plants is very small and often undetectable.

About thirty species of grasses are now known which have nitrogen-fixing bacteria associated with their roots, including varieties of maize (one instance of which pre-dated the excitement about maize and *S. lipoferum*). They have been detected by the acetylene test but not studied in detail—some are 'couch grasses' found even in British gardens (e.g. *Cynodon dactylon*). Some temperate weeds such as hedge woundwort and the persistent ground elder also give positive acetylene tests. In all these instances we do not yet know whether the association is casual (involving no lasting interdependence—see section 5.5) or a true associative symbiosis. But this type of association has only been known for a few years and might yet prove to be of great economic importance. The disappointing performance of *S. lipoferum* with maize was really the deserved failure of a shot in the dark; it is early days yet, and with more detailed knowledge of the relative roles of plant and microbe it may be quite easy to develop exploitable nitrogen-fixing associations involving cereal crops.

5.5   Casual associations

There are a number of instances in which nitrogen-fixing bacteria associate with higher organisms and in which the benefit incurred by one partner in the association is small or negligible. These will be discussed briefly in this section.

5.5.1   Leaf associations

The association of blue-green algae with the bases of leaves *(Gunnera)* or leaf-like organs *(Azolla,* liverworts) are true symbioses. More casual associations occur when nitrogen-fixing bacteria grow on the leaves of plants, then become washed into soil by rain and thereby up-grade the nitrogen status of the soil near the plant roots. The surface zone of the leaf where microbes grow (such as fungi or bacteria able to utilize leaf exudates) is called the 'phyllosphere'. In many plants of the wet tropics, such as cocoa and sugar cane, the phyllosphere is populated by *Beijerinckia* which very probably up-grade the local soil after tropical rainstorms. In other instances—fir trees in temperate regions, for example—it is very doubtful whether even a casual association really exists. Much of the research in this area has not taken account of the ease with which ghosts (section 4.5) can be isolated from such habitats.

A more sophisticated leaf association, which was thought for some years to be a nitrogen-fixing symbiosis, is the leaf nodule system. Certain tropical plants *(Ardesia, Psychotria*—the latter is quite a popular house plant in Britain) have tiny nodules on their leaves which, as has been known for several decades, are colonized by bacteria with consequent benefit to the plant. Among the bacteria readily isolated from such nodules are nitrogen-fixing klebsiellae (see Table 5) so, in the early 1960s, the idea that the beneficial effect on the plant was due to fixation of nitrogen was plausible. Since then the idea has had to be discarded, for two main reasons:

(1) The beneficial bacteria are by no means always able to fix nitrogen.
(2) Even if they can do so, the acetylene test indicates that they cease fixing nitrogen when they colonize the leaf nodule.

It seems probable that the benefit to the plant arises from the formation by the bacteria of hormone-like substances in the leaf nodule, not from nitrogen fixation.

5.5.2   Root associations

Free-living nitrogen-fixing bacteria are often found round the roots of
plants, no doubt living on root secretions and benefiting themselves
rather than the plant. Mycorrhiza, a group of symbiotic fungi, associates
symbiotically with many types of plant, assisting uptake of nutrient (such
as phosphates) and enhancing vigour in ways which are not often
understood. In the mid 1960s there seemed to be some evidence that the
mycorrhiza of a gymnosperm, *Podocarpus*, fixed nitrogen but more recent
evidence suggests that the mycorrhiza simply supply a favourable
environment for fixation by ordinary soil bacteria such as klebsiellae and
bacilli. Indeed casual associations between enterobacteria or bacilli and
roots seem quite common and the borderline between these and
associative symbioses is difficult to define. In principle, a symbiosis
benefits both partners directly and a casual association benefits only one,
but indirect benefit to the second partner can obviously arise in many
associations which seem at first sight to be casual.

5.5.3   Destructive associations

By altering the nitrogen status of the local environment, free-living
nitrogen-fixing bacteria can enhance microbial degradation and
deterioration processes. They do so in compost heaps, decaying leaf litter
and other natural environments where the C : N ratio is high. A clear-cut
case occurs in the rotting timber: nitrogen-fixing coliform bacteria (see
section 4.2) can multiply alongside wood-rotting fungi and accelerate the
rotting of wood by making fixed nitrogen available to the fungus. Here
the bacteria benefit from being able to utilize the products of fungal
degradation of cellulose.

5.5.4   Associations with animals

Termites and other wood-eating insects have nitrogen-fixing
citrobacters (see section 4.2) in their intestines and, when the host is on a
low nitrogen diet, they actually fix some nitrogen. The acetylene test
indicates that their contribution to the nitrogen status of the insect is, at
best, trivial. Butyric clostridia (see section 4.1) are normally found in the
rumens of sheep and cows but their nitrogen-fixing activity is slight and
could not, at best, account for more than 1% of the animal's N turnover.
They have probably been ingested with food; this is certainly the case with
winter reindeer which, after feeding on reindeer moss, can have a lot of
nitrogenase activity in their rumen due to the phycobiont (see section 5.3).

Klebsiellae capable of nitrogen fixation have been isolated from
human intestines but, again, it is unlikely that they ever contribute serious
amounts of N to their host. Generally speaking, the animal intestinal tract
contains sufficient free ammonia to repress (see section 3.5) any serious
potential nitrogen-fixing activity.

6 Genetics and Evolution

The study of the hereditary information which enables organisms to fix nitrogen is somewhat simplified by the fact that the property is restricted to non-nucleated microbes; the bacteria and blue-green algae. Some of the complications of the genetics of higher plants and animals are thus avoided: division of genetic information among many chromosomes, duplication of genetic information, sexual exchange of genomes, rarely present problems with bacteria. The bacterial chromosome is a circle of DNA (deoxyribonucleic acid) consisting of two intertwined circular molecules between which all the genetic information of the organism is divided. Many bacteria carry one (or sometimes more) additional DNA circles called plasmids; these are less than 1% of the size of the chromosome and supply supplementary information conferring properties such as resistance to drugs, ability to transfer DNA to other bacteria or ability to utilize certain substrates. They will be discussed further in section 6.1.2.

The techniques for studying microbial genetics are relatively simple, but a detailed account would not be appropriate here and for further information the reader should consult a textbook of microbial genetics. It is worth recalling, however, that genetic information is encoded in fragments of the DNA called genes, the code being based on the detailed chemical structure of that portion of the DNA chain. One gene generally codes for one peptide molecule (peptides are the fundamental chains of amino acids of which proteins are composed; some proteins consist of only one peptide). A change in the chemical structure anywhere along the DNA constituting a certain gene will cause mis-reading: either the wrong peptide will be formed or none at all. Such a change is called a mutation. Mutations occur spontaneously in about one in 10 to 100 million of a normal bacterial population; their frequency can be increased considerably by treatment of the microbes with certain chemicals or with ionizing radiation. The study of mutants is basic to microbial genetics; not only are mutations invaluable markers for identifying genetic types of microbe, they can also be used to obtain maps of chromosomes, to study the regulation of genes (i.e. the control of the extent to which genetic information is used by the cell) and even to generate organisms with entirely new properties. In this chapter the application of modern methods of microbial genetics to the study of nitrogen fixation will be discussed briefly.

6.1   The nitrogen fixation gene cluster

The genetic information enabling bacteria to fix nitrogen is given the shorthand name *nif* by microbial geneticists. At this stage it will be useful to make a piece of conventional nomenclature clear. For convenience of discussion, all microbial genes have shorthand, three-letter names of this kind. Any mutation in the gene which leads to failure of its expression is given a minus sign: a mutant with mutated *nif*, which can no longer fix nitrogen, is called *nif⁻*. However, mutations in certain other genes can cause failure to fix nitrogen so, if a mutant is obtained which cannot fix nitrogen but which has not been positively identified as mutated in *nif*, it is termed Nif⁻ (a capital letter and no italics). Correspondingly, if a mutant which is *nif⁻* is treated genetically to restore its nitrogen-fixing capacity, yet there is uncertainty whether the original *nif⁻* mutation has been reversed, it is called Nif⁺. In more technical terms, *nif⁺* and *nif⁻* refer to wild type and mutant genes specifically; Nif⁺ and Nif⁻ refer to ability or otherwise of mutants to fix nitrogen, whether or not mutations in the *nif* genes are responsible.

Though Nif⁻ mutants of *Azotobacter vinelandii* and *Clostridium pasteurianum* had been reported in the 1960s, study of the genetics of *nif* really got started with some gene transfer experiments early in the 1970s. *nif⁻* mutants of *Klebsiella pneumoniae* were obtained and, by conventional methods of gene transfer, *nif* from wild type *K. pneumoniae* was introduced into the *nif⁻* mutants and their mutations corrected. From being Nif⁻ they became Nif⁺. The two procedures used to make those gene transfers are *transduction* and *conjugation*.

6.1.1   Transduction

Bacterial viruses are called bacteriophages. Certain of them, when they infect a new host, can transfer some of the genetic material of the old host with them. The co-transfer process is called transduction. If the new host survives the virus infection, and a few cells always do, it now possesses some genetical material from the virus's previous host. To transfer *nif* to *nif⁻* recipients, a wild type *nif⁺* *K. pneumoniae* was infected with a bacteriophage called P1, which was known to be good at transduction. The virus was recovered from the infected wild type and used to infect *nif⁻* mutants. A proportion of these became Nif⁺, indicating that sometimes *nif* genes from the wild type had co-transferred with P1.

6.1.2   Conjugation

Certain of the plasmids mentioned at the opening of the chapter confer upon their host the ability to donate the plasmid to a new host by a sort of pseudo-sexual process called conjugation. The donor cell carrying the plasmid (the male) actually comes alongside the recipient (female) organism, a tube grows between them and a replica of the plasmid forms and is transferred to the recipient. This, therefore, becomes a male and can donate the plasmid to yet another recipient. Not all plasmids, by any means, have this so called 'sex factor activity', but those that do can sometimes co-transfer fragments of the donor's chromosomal DNA to the recipient, just as the bacteriophage did in transduction. To transfer *nif* to *nif⁻* recipients, a plasmid carrying both sex factor activity and resistance to the antibiotic Kanamycin was introduced from its natural host, *Escherichia coli*, into wild type *K. pneumoniae*. The *K. pneumoniae* then became 'male'. It was 'mated' with *nif⁻* mutants of *Klebsiella* and the progeny acquired resistance to Kanamycin, as would be expected when the mutants received a replica of the plasmid. A small proportion of these had also become Nif⁺, showing that the plasmid had, at low frequency, mobilized some *nif* genes from its host's chromosome and co-transferred them to the mutants.

6.2   The genetic map of *nif*

Gene transfer experiments of the kind just described can give one information about the location of genes on the chromosomes. For example, if the *nif⁻* recipient of *nif* also had a mutation in a gene cluster concerned with making the amino acid histidine (a gene cluster called *his*), *his⁻* and *nif⁻* mutations were corrected simultaneously fairly often in gene transfer experiments. But if the *nif⁻* recipient also had a mutation in a gene cluster *(trp)* concerned with making the amino acid tryptophan, the simultaneous correction of *trp⁻* and *nif⁻* was an extremely rare event. It followed that *nif* lies fairly close to *his* on the *Klebsiella* chromosome and a long way from *trp*. Numerous experiments of this kind have converged to give the coarse 'map' in Fig. 6–1; the genes closest to *nif* are *his* on one side and *shi* on the other.

It is also possible to use gene transfer experiments to obtain a finer map. The *nif* cluster must consist of several genes: at least two for the different polypeptides of the Mo–Fe protein and one for the Fe protein (see Table 2) as well as at least one to regulate formation of enzyme according to whether ammonia is present or not (see section 3.5). Mutants in *nif* are now known which lack the Mo–Fe protein, the Fe protein, or both; by transducing these *nif⁻* mutations from one to another, always in a klebsiella with a *his⁻* mutation, one can arrange them in order relative to the *his⁻* mutation and to each other. Such transductional mapping has indicated eight distinguishable genes in the *nif* cluster already; they are illustrated in the map of *nif* in Fig. 6–2, though the order of *nif* K, D and H is not definitely established at the time of writing. Fig. 6–2 also indicates what the genes code for (as far as this is known).

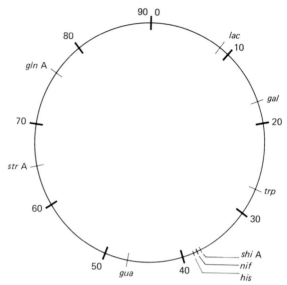

Fig. 6–1 A stylized map of *Klebsiella*'s chromosome showing the relative position of the nitrogen fixation genes. Several of the other genes quoted are positioned by analogy to the closely related *Escherichia coli* because their location in *Klebsiella* is not certain. The numbers are relative distances out of 90 units from the point at which replication is initiated (o).

lac   genes for a lactose-utilizing enzyme.
gal   genes for a galactose-utilizing enzyme.
trp   genes for synthesis of tryptophan.
shi A   a gene concerned in uptake of shikemic acid.
nif   genes for nitrogen fixation.
his   genes for synthesis of histidine.
gua   genes for synthesis of guanine.
str A   a gene conferring resistance to the antibiotic streptomycin.
gln A   a gene for synthesis of glutamine synthetase.

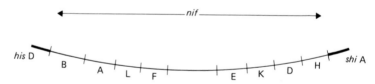

Fig. 6–2 Finer map of *Klebsiella*'s nitrogen fixation genes. Symbols as for Fig. 6–1. Capital letters indicate single genes within the *nif* gene cluster.

K and D   code for the molybdoprotein of nitrogenase.
H   codes for the iron-protein.
E, A and L   are regulatory sites.
F   codes for something concerned with reduction of nitrogenase.
B   codes for a part of the molybdoprotein carrying molybdenum.

6.3   Genetic regulation of *nif*

The *nif* cluster can be seen from Fig. 6–2 to include three genes (A, L, E) which 'switch off' synthesis of nitrogenase. Mutations in these genes prevent synthesis of both the Mo–Fe protein and the Fe protein (specified by D, K and H) together. These are called 'regulatory' genes; they either form or recognize substances which regulate the reading of 'structural' genes, which specify the enzyme polypeptides. There exist a class of mutants which are Nif⁻ but which have a mutation which does not map in *nif* at all; the mutations map far away on the chromosome at a site called *gln* (see Fig. 6–1). This site carries genes coding for the enzyme called glutamine synthetase (see section 1.3), an important substance in general cell metabolism. To simplify a complex story, involving a lot of research, we now know that glutamine synthetase can exist in active and inactive forms as far as enzyme activity is concerned, and that the presence of ammonia causes the microbe to perform a sequence of reactions whereby the enzyme becomes inactive. Now, the active form is not only an enzyme; it is also a regulator substance capable of causing the microbe to start using several gene clusters concerned with nitrogen metabolism. Examples are the utilization of urea, nitrate or certain amino acids. Looked at teleologically, the deficiency of ammonia causes the microbe, by way of glutamine synthetase, to 'switch on' its information telling it how to mobilize other sources of nitrogen which might be in its environment. Active glutamine synthetase also switches on the *nif* genes, enabling the possessors of those genes to use atmospheric nitrogen if all else fails. Which site on *nif* recognizes glutamine synthetase is not known, and, indeed, the chances are that switch-on of *nif* occurs in more than one stage (i.e. that a second regulator substance is formed in response to the recognition of active glutamine synthetase), but these are complications which are still being studied. One interesting experiment has been done with klebsiella, however. It is possible to mutate the *gln* region and obtain mutant microbes which always make active glutamine synthetase, even if ammonia is present. If these possess *nif* genes, then they make nitrogenase (and fix nitrogen) even if ammonia is present. They are called *constitutive* Nif⁺ strains.

6.4   *Nif* plasmids

In section 6.1 I described how a plasmid was used to mobilize the *nif* genes from the klebsiella chromosome and transfer it to *nif⁻* mutants. I also emphasized the closeness of *nif* genes to *his*. Since the original plasmid came from *E. coli*, it would transfer back to its original host readily and one might hope that, albeit at low frequency, it might co-transfer *nif* genes into the organism. If the *E. coli* could use the *nif* genes, an entirely new species of nitrogen-fixing bacteria would thus have been generated. The writer's colleagues attempted this experiment early in the 1970s. In outline, a specially chosen *E. coli* strain with a *his⁻* mutation was mated with the 'male' nitrogen-fixing klebsiella described in 6.1 and His⁺ progeny were obtained at a low frequency. Nearly all of these, to our delight, were able to fix nitrogen: the *nif* genes had co-transferred with *his* and the *E. coli* was able to use them. Ammonia repressed *nif*, so the *E. coli* had the right sort of glutamine synthetase to 'switch on' *nif*.

A close examination of the nitrogen-fixing *E. coli* hybrids showed that two classes of event had occurred:

(1) A strain had arisen in which *his* and *nif* from klebsiella had integrated completely into the *E. coli* chromosome. It was genetically a new type of nitrogen-fixing organism.

(2) Two strains had arisen in which the klebsiella DNA had not integrated into the chromosome, but had formed new plasmids; not just one, but two in one case and three in another. The *his* and *nif* genes were on one of these.

The spontaneous formation of plasmids carrying *nif* in *E. coli* lent credibility to another idea for genetic manipulation of *nif*: could one not build *nif* into a plasmid which had its own sex factor activity? If so, *nif* would transfer at every mating, not just co-transfer (with luck) in about 1% of matings. There exist established techniques for constructing such plasmids, for which the specialized literature should be consulted; their application led to the successful construction, by one of the author's associates, of three different *nif* plasmids; maps of the two most useful ones are sketched in Fig. 6–3. Both carry drug resistance genes which are useful in the laboratory. They have been used for a variety of research purposes and several new types of nitrogen-fixing bacteria have been constructed with them. Among important consequences of the use of such plasmids have been the following: evidence for the role of glutamine synthetase in regulation of *nif*; evidence for a molybdenum uptake pathway shared with another enzyme; information on the stability of klebsiella's *his–nif* genes in alien genetic backgrounds. It is a curious thought that, since these plasmids were constructed in *E. coli*, that organism is their 'natural' host and much of the genetics of klebsiella *nif* is now studied in *E. coli*! But some return traffic is taking place: one can mutate *nif* on the plasmids and, by transferring them back to *nif⁻* mutants of klebsiella, obtain independent confirmation of the genetic map in Fig. 6–2.

6.5   New nitrogen-fixing bacteria

The nitrogen-fixing *E. coli* hybrids described in the previous section were the first types of new nitrogen-fixing microbes to be developed

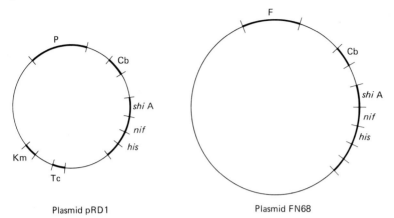

Fig. 6–3   Speculative maps of the plasmids carrying nitrogen fixation genes. The relative positions of the genes are not established except for those including *his* and *nif* from *Klebsiella*. Abbreviations as for Fig. 6–1. Additional abbreviations:

F and P   genes for sex factor activity.
Cb   gene conferring resistance to carbenicillin.
Tc   gene conferring resistance to tetracycline.
Km   gene conferring resistance to Kanamycin.

deliberately by genetic manipulation. This success opened up some of the future possibilities discussed in Chapter 7. Numerous transfers of *K. pneumoniae nif* have now been made using such plasmids, some of which are illustrated in Fig. 6–4.

Bacteria such as *E. coli, Salmonella typhimurium* or *Klebsiella aerogenes* become perfectly effective nitrogen-fixing bacteria when provided with *nif*, though they still only fix anaerobically. Equally interesting are organisms such as *Agrobacterium tumefaciens*, which does not fix nitrogen even if it possesses klebsiella's *nif* genes (one can show that recipients of the plasmid have *nif*, because they become 'male' and will transfer the plasmid onwards to *E. coli*, which then fix nitrogen quite readily). Evidently *A. tumefaciens* lacks some supplementary information which is necessary for complete expression of *nif*. Yet another interesting transfer is that in which a plasmid carrying klebsiella *nif* was used to correct *nif*⁻ mutants of *Azotobacter vinelandii*. These were known to have mutations in genes specifying the Mo–Fe protein in one mutant and the Fe protein in another, so the Nif⁺ progeny must have been using klebsiella-type proteins for at least part of their nitrogenase. Not only did azotobacter's regulatory system, its glutamine synthetase perhaps, serve to switch on klebsiella *nif*: the corrected mutants could, in fact, grow and fix nitrogen in air, which klebsiella cannot normally do, so klebsiella nitrogenase in azotobacter can share its new host's oxygen-excluding stratagems (see section 3.1).

Reports have appeared in the scientific literature of the transfer of *nif* genes from *Rhizobium trifolii* to a strain of *Klebsiella aerogenes* which does not normally fix nitrogen, and also of the transfer of *nif* genes between photosynthetic bacteria. The latter reports are of a very preliminary nature.

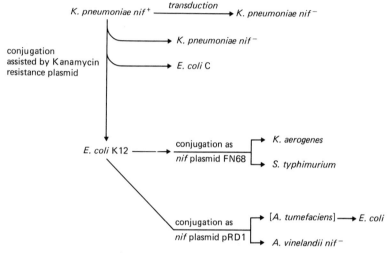

Fig. 6–4   Some transfers of *Klebsiella pneumoniae* nitrogen fixation genes to new bacterial hosts. For details see section 6.1.1, 6.1.2 and 6.4.

6.6   Artificial *nif* plasmids

In the mid 1970s, techniques were developed for genetic manipulation in the test tube: DNA could be isolated from a source, cut into pieces (figuratively speaking) with special enzymes (restriction enzymes) and re-assembled with other enzymes (ligases). If DNA from different organisms were cut, mixed and re-assembled, all sorts of mixed genetic elements could be constructed artificially. If, finally, part of the DNA came from an identifiable genetic unit such as a bacteriophage (see 6.1.1) or a plasmid, one could construct bacteriophages or plasmids carrying alien genetic information and these would multiply in a host such as *E. coli*, producing numerous copies of themselves and therefore of the new DNA. Under the name of 'genetic engineering' (a name which had been used before for more conventional genetic manipulations) this technology achieved brief notoriety because, by careless insertion of alien genes (e.g. ability to cause disease or cancer) into the genetic complement of *E. coli*, one might make it into a really nasty microbe. Yet one could equally introduce highly desirable properties, such as ability to make rare hormones, into bacteria, which could then be grown in large quantities. In the present context, small plasmids or bacteriophages carrying the *nif* cluster would be much more concentrated sources of *nif* DNA for general experimental work—including transfer to exotic hosts. At the time of writing (early 1977), the *nif* gene cluster has proved amenable to such artificial manipulation but it is cut somewhere in the middle by those restriction enzymes used so far. The left hand section of the map in Fig. 6–2 (*nif* B, A, F, and L, together with part of *his*) has been spliced into a small plasmid which has the property, in special circumstances, of forming several hundred copies of itself per bacterium. Thus great quantities of *nif* DNA in a relatively purified form have become available for research. The rest of the *nif* cluster will no doubt be 'cloned' in this way fairly soon.

6.7   Evolution of nitrogen fixation

It is worth stepping back from genetics for a moment to consider the implications of recent biochemical and genetical findings on the origin and evolution of nitrogen fixation. Though there is little of a factual nature to go on, some instructive speculations are possible. For example, for many years scientists were generally of the view that nitrogen fixation was an ancient property, largely because it is only found among the most primitive of living things—the bacteria—and even among those it seemed most prevalent among the more primitive types, the anaerobes. If it was an ancient property, it must be one which was readily lost during the development of higher organisms. Some support for this view comes from the detection of traces of what may be blue-green algae with heterocysts in very ancient pre-Cambrian rocks (over 3×10^9 years old). On the other hand, what good would nitrogenase have been in those days when, according to geochemical dogma, the planetary atmosphere and environment contained no oxygen but some free ammonia? Could the enzyme in fact have been doing something quite different, such as removing toxic cyanides? Yet if the enzyme is so old, why has there been so little evolutionary divergence in its structure? It is much the same wherever it comes from, even to the extent of giving the cross-reactions described in section 2.1. And why have no plants or fungi learned how to use it? After all, the properties of blue-green algae or azotobacter tell us that there is no serious obstacle to nitrogen fixation by ordinary aerobic creatures. An alternative view is that it is not an ancient property. The ability to fix nitrogen may have emerged late in evolutionary history: sometime after oxygen became a permanent component of the planetary atmosphere and after plants had used up all superficial supplies of ammonia, nitrates and other sources of fixed N. If it emerged once, in a bacterium, then we now know that it could spread quite easily among other bacteria—perhaps even aided by plasmids, as in the laboratory —and this would account for the scattered way in which *nif* is today distributed throughout the microbial world. It would also account

7 The Future

The importance of the nitrogen cycle in the productivity of the biosphere has caused many people, not only scientists, to focus their attention on nitrogen fixation, its rate-determining step (see section 1.1). In sections 1.7 and 1.8 some indication was given of the present situation regarding nitrogen fertilizer and world food supplies; essentially, some dramatic changes in agricultural practice will be needed by the twenty-first century if adequate nitrogen is to be available to feed the world's expanding population, even if all current birth control programmes are reasonably successful. Much improvement is already possible from the application of existing knowledge such as that outlined in earlier chapters, but the probability is that these will only give us breathing space to develop longer term solutions. The types of approach which are possible are briefly discussed below.

7.1   Chemical fertilizers

Chemical fertilizers have a number of defects; they need a sophisticated industry, they cost a lot in terms of energy, they have high transport costs, they are wasteful and often pollute the environment. Yet they work and there is no getting away from this fact. If they could be produced more cheaply, or be produced by simple means appropriate to developing countries, and if their use could be controlled intelligently, then the future of agriculture would look brighter. Recent developments in the chemistry of nitrogen have opened several possibilities for modifying the traditional Haber process or replacing it by altogether simpler procedures.

7.2   Botanical

The useful agricultural nitrogen fixers are the legumes. Plant physiologists have recently developed a non-sexual method of obtaining hybrids which is called 'somatic cell hybridization'. Suitably treated plant cells (e.g. from a leaf) can be induced to fuse in pairs and the products can sometimes be regenerated into whole new plants (hybrid petunias have been prepared in this way). Legumes do not hybridize sexually with, for example, brassicas, but somatic cell hybridization might yield such hybrids and thus confer the property of nitrogen fixation on some large edible green vegetable. Perhaps more seriously, there is some prospect of preparing, by somatic hybridization, a cereal which forms nitrogen-fixing nodules like those of a soya bean. The prospect is still remote, but the cells of barley and soya bean have in fact been seen to fuse in appropriate conditions.

7.3   Genetical

The transfer of bacterial *nif* genes to a plant would not, in itself, give those plants the ability to fix nitrogen. (Indeed, given the intimacy of many nitrogen-fixing plant-microbe associations, it is probable that *nif* genes enter plant cells fairly often in nature when senescence or damage occurs.) Yet there are all sorts of ways in which the plant might be 'taught' to fix nitrogen. The *nif* genes could be introduced on a plant virus, together with supporting genes for regulating *nif*, for keeping oxygen out of the way and so on. The necessary genes for nitrogen fixation might be 'spliced' into some kinds of plant DNA by artificial means in a test tube. The *nif* genes might be introduced into benign parasitic bacteria or fungi which would persist inside or close to the plant.

One might even by-pass plants and put *nif* genes into the sorts of bacteria which are symbiotic with animals: a nitrogen-fixing goat (its rumen populated with nitrogen-fixing bacteria) might flourish on a diet of paper even better than normal goats are reputed to do! Or one might use the microbes as little ammonia factories: make a constitutive Nif$^+$ strain (section 6.3) of a blue-green alga, introduce another mutation so that it cannot use the ammonia it forms (such mutants have already been made in klebsiella) and you have an organism that makes nitrogen into ammonia at the expense of solar energy!

7.4   Consequences

There are numerous bright ideas one may have for the future and they all have complications when one considers them in detail. But many are plausible in principle; the specialized literature should be consulted for amplification of some of those just mentioned. If, by some means or another, a successful break-through is achieved and (for example) a natural nitrogen-fixing cereal is developed, would it be wholly beneficial? Some environmentalists have expressed alarm: fears of rampant nitrogen-fixing weeds upsetting the world's ecological balance, for example. It is always wise to consider the possible risks of scientific innovation, and there are indeed risks in the indiscriminate spread of nitrogen fixation. But on consideration they prove to be minor ones, and the potential benefits are vast. Consider what would happen if nitrogen-fixing plants including cereals became a reality. Essentially, they could behave rather like soya beans do today: nitrogenous matter would cease to limit their productivity and other minerals such as potassium, sulphur or phosphate would become limiting. But the manufacturing and transport costs of nitrogenous fertilizers would have been saved, as well as much of the environmental cost due to 'run-off' of unused fertilizer into lakes and drinking water. If, by careful agricultural management, adequate K, S, P, etc., were supplied, then carbon dioxide would become the limiting nutrient: as with soya beans in glasshouse conditions today, photo-assimilation of CO_2 would limit productivity. (Intensive farmers would probably spend their money on enriching the atmosphere of huge plastic greenhouses with CO_2.) Clearly the prospect of a catastrophic environmental 'take-over' by nitrogen-fixing weeds and grasses is not to be taken seriously (it would probably have happened already if it were feasible). But in this life one does not get something for nothing: a nitrogen-fixing cereal would grow rather more slowly than its cousin being supplied with ammonia, because some of its ATP would be needed to make nitrogenase work (see section 2.2). Looked at from another view-point, the northernmost limits of a nitrogen-fixing cereal would be a little further south than those of its cousins growing with ammonia. This price would be small compared with the savings in costs of artificial fertilization.

There are also some minor hazards to be considered, though, once pathogenic bacteria operate by rotting the living plant, a slow procedure because plants are generally not rich in fixed N. If such microbes picked up *nif* genes and could use them, they would become more pathogenic. Another hazard: ordinary lakes and rivers have very little fixed N. If one of the more obstinate water weeds acquired the ability to fix nitrogen, it might well clog up waterways and reservoirs seriously, just as nitrogen-fixing algae occasionally 'bloom' in waters contaminated with phosphates and cause expensive pollution problems. No doubt yet other local risks could be foreseen, but they fade into insignificance when considered in relation to the possibility of feeding a world population of over 7×10^9 people: this is what is promised by the judicious spread of really effective nitrogen-fixing ability to cereals such as rice, maize, wheat and barley.

Bibliography

Books

BURNS, R. C. and HARDY, R. W. F. (1975). *Nitrogen Fixation in Bacteria and Higher Plants*. Springer, New York.

MISHUSTIN, E. N. and SHILNIKOVA, V. K. (1968). *Biological Fixation of Atmospheric Nitrogen*. English translation 1971, Macmillan, London.

POSTGATE, J. R. (ed.). (1971). *The Chemistry and Biochemistry of Nitrogen Fixation*. Plenum, New York.

QUISPEL, A. (ed.). (1974). *The Biology of Nitrogen Fixation*. North Holland, Amsterdam.

STEWART, W. D. P. (1966). *Nitrogen Fixation in Plants*. Athlone Press, London.

Practical

ANON. (1967). *A Practical Manual of Soil Microbiology Methods*. Soils Bulletin No. 7. Food and Agricultural Organization, Rome.

VINCENT, J. M. (1970). *A manual for the practical study of root nodule bacteria* (IBP Handbook 5). Blackwell, Oxford.

Symposia

NEWTON, W. E. and RYMAN, C. J. (eds.). (1976). *Proceedings of the 1st International Symposium on Nitrogen Fixation*. Vols. 1 and 2. Washington State University Press, Pullman.

NEWTON, W. E., POSTGATE, J. R. and RODRIGUEZ-BARRUECO, C. (eds.). (1977). *Recent Developments in Nitrogen Fixation. Proceedings of the 2nd International Symposium on Nitrogen Fixation*. Academic Press, London.

NUTMAN, P. S. (ed.). (1976). *Symbiotic Nitrogen Fixation in Plants*. (IBP7). Cambridge University Press, London.

STEWART, W. D. P. (ed.). (1976). *Nitrogen Fixation by Free-living Organisms*. (IBP6). Cambridge University Press, London.

Reviews

BRILL, W. J. (1975). *Ann. Rev. Microbiol.*, **29**, 109–30.

DILWORTH, M. J. (1974). *Ann. Rev. Pl. Physiol.*, **25**, 81–114.

WINTER, H. C. and BURRIS, R. H. (1976). *Ann. Rev. Biochem.*, **45**, 409–26.

YATES, M. G. and JONES, C. W. (1974). *Adv. Microb. Physiol.*, **11**, 97–136.

ZUMFT, W. G. and MORTENSEN, L. E. (1975). *Biochim. biophys. Acta.*, **416**, 1–52.

Popular

VARIOUS AUTHORS (1977). Nitrogen—a special issue. *Ambio* (Sweden), **6**, parts 2–3.